Collins

Cambridge IGCSE™

ICT

TEACHER'S GUIDE

Paul Clowrey

William Collins' dream of knowledge for all began with the publication of his first book in 1819.
A self-educated mill worker, he not only enriched millions of lives, but also founded a flourishing publishing house. Today, staying true to this spirit, Collins books are packed with inspiration, innovation and practical expertise.
They place you at the centre of a world of possibility and give you exactly what you need to explore it.

Collins. Freedom to teach.

Published by Collins
An imprint of HarperCollins*Publishers*
The News Building, 1 London Bridge Street, London, SE1 9GF, UK

HarperCollins Publishers
1st Floor, Watermarque Building, Ringsend Road, Dublin 4, Ireland

Browse the complete Collins catalogue at
www.collins.co.uk

© HarperCollins*Publishers* Limited 2022

10 9 8 7 6 5 4 3 2

ISBN 978-0-00-843093-1

All rights reserved. No part of this publication may be reproduced, stored in a retrieval system, or transmitted in
any form by any means, electronic, mechanical, photocopying, recording or otherwise, without the prior written permission
of the Publisher or a licence permitting restricted copying in the United Kingdom issued by the
Copyright Licensing Agency Ltd, 5th Floor, Shackleton House, 4 Battle Bridge Lane, London SE1 2HX.

British Library Cataloguing-in-Publication Data
A catalogue record for this publication is available from the British Library.

Author: Paul Clowrey
Previous edition author: Colin Stobart
Publisher: Elaine Higgleton
Product manager: Alex Marson
Proofreader: Alison Bewsher
Cover designer: Gordon MacGilp
Typesetter: Jouve UK LTD
Production controller: Lyndsey Rogers

Cambridge International copyright material in this publication is reproduced under licence and remains the intellectual property of Cambridge Assessment International Education.

Third-party websites and resources referred to in this publication have not been endorsed by Cambridge Assessment International Education.

Printed and bound in the UK using 100% Renewable Electricity at CPI Group (UK) Ltd

MIX
Paper from responsible sources
FSC® C007454

This book is produced from independently certified
FSC™ paper to ensure responsible forest management.

For more information visit:
www.harpercollins.co.uk/green
Acknowledgements

The publishers gratefully acknowledge the permission
granted to reproduce the copyright material in this book.
Every effort has been made to trace copyright holders
and to obtain their permission for the use of copyright
material. The publishers will gladly receive any
information enabling them to rectify any error or omission
at the first opportunity.

Contents

Introduction	iv
Two-year scheme of work	vi
Syllabus coverage table	x

LESSON PLANS — 1

UNIT 1	Types and components of computer systems	2
UNIT 2	Input and output devices	26
UNIT 3	Storage devices and media	46
UNIT 4	Networks and the effects of using them	64
UNIT 5	The effects of using ICT	88
UNIT 6	Data types and databases	122
UNIT 7	Systems analysis and design	146
UNIT 8	Safety and security	164

Introduction

Welcome to the Collins Cambridge IGCSE™ ICT Teacher's Guide, which we hope will provide invaluable support to teachers worldwide as they support learners who are following the Cambridge IGCSE and IGCSE (9-1) Information and Communication Technology syllabuses (0417/0983) for examination from 2023.

Using the Student's Book

The Student's Book offers an integrated approach to the Cambridge Information and Communication Technology syllabus. It is divided into eight main units, each of which links one main theory section with practical sections of the syllabus.

A breakdown of the units in the Student's Book, and the syllabus sections each one covers, can be found on pages 8 and 9 of this Teacher's Guide.

Units and sessions in the Student's Book

Each unit in the Student's Book comprises sessions that cover the theory (for example, in Unit 1, *The main components of a computer*) followed by sessions that cover the practical skills (for example, again in Unit 1, *Creating a newsletter*). Each unit starts with an introduction to the topic, and outlines a practical task that makes a connection between what students will learn in the theory session and the skills that they will be developing in the practical sessions.

Active learning

As well as offering an integrated approach, the Student's Book promotes active learning on the part of the student, in terms of a range of activities that emphasise searching, collating and presenting, as well as developing practical IT skills. Answers to all these Student's Book activities are contained in this Teacher's Guide, in the lesson plans that accompany each session in the Student's Book.

Practice questions

Each Student's Book unit contains a theory review section at the end of theory sessions and a practical review section at the end of practical sessions. All these review sections contain practice questions to test what the students have just learned in the unit. Answers to all these review questions can be found in the accompanying digital download.

Student's Book digital download

The Student's Book digital download contains the following three elements:

- source files for use with specific practical tasks in the Student's Book units
- practice exam-style question papers and answers
- extra support containing information sheets on creating websites and using spreadsheets.

Using the Teacher's Guide

The Teacher's Guide comprises two components – the printed book and a digital download. The **printed book** contains:

- lesson plans to accompany every session in the Student's Book
- worksheets to supplement and extend activities in the Student's Book
- a suggested two-year scheme of work
- a syllabus coverage table

The **digital download** contains:
- all lesson plans and worksheets in Word format (as well as in PDF format) so that they can be edited to suit the needs of individual classes or departmental schemes of work
- suggested solutions to worksheet questions and tasks (where appropriate)
- additional source files for use with specific practical tasks
- PowerPoint slides to supplement and extend Student's Book activities or exemplify concepts
- answers to the theory and practical review sections in the Student's Book
- the suggested scheme of work as an editable Word file

Lesson plans

Every session in the Student's Book is intended to provide work for one or two lessons and is supported in the Teacher's Guide by a two-page lesson plan. Each lesson plan follows the same format, making it easy to use and prepare lessons.

The following sections of each lesson plan help in preparing lessons:
- The session summary highlights the main ideas or aspects covered by the lesson.
- Differentiated learning outcomes indicate what students should master during the lesson.
- Resources highlight what you need to deliver the lesson, including Student's Book page numbers, source files from the Student's Book digital download or the Teacher's Guide digital download, worksheets, PowerPoints, equipment if necessary, and suggestions for websites or examples of software.
- These elements are followed by suggestions for the structure of the lesson:
- Starter suggestions involve the whole class and give you ideas on how to capture the attention and interest of students.
- Main lesson activities help you lead students into activity questions in the Student's Book or on the worksheets.
- Plenary suggestions offer guidance on how to round off lessons.

Finally, at the end of each lesson plan, you will find suggestions for assessment and/or homework.

Two-year scheme of work

The tables in these pages suggest a week-by-week plan for full coverage of the syllabus over a two-year course. Each week, generally, comprises one theory and one practical lesson, with each lesson being a session from the Student's Book. There are 15 sessions (identified by *italics*) that are more comfortably covered in two lessons.

The syllabus is designed for around 130 hours of guided learning, but teachers should bear in mind that this number of hours may vary according to individual school or college practice and students' prior experience or subject knowledge.

Year 1 – covering Units 1, 2, 3 and 4 with consolidation and revision:
35 weeks – 70 hours * T(heory)/P(ractical)

Week	Session	T/P*
Unit 1		
1	1.1 Types of computers	T
	1.6 Document production	P
2	1.2 Hardware and software	T
	1.7 Internet research	P
3	1.3 The main components of a computer	T
	1.8 Creating a newsletter	P
4	1.4 Operating systems	T
	1.9 Email	P
5	*1.5 Using a computer system to communicate*	T
6	Consolidate and review all Unit 1 sessions	T & P
7	Unit 1 Theory review	T
	Unit 1 Practical review	P
Unit 2		
8	*2.1 Input devices*	T
	2.5 An introduction to presentations	P
9	*2.1 Input devices*	T
	2.6 Presentation preparation	P
10	2.2 Direct data entry devices	T
	2.6 Presentation preparation	P
11	*2.3 Output devices*	T
	2.7 Adding text, images and multimedia elements to a presentation	P
12	*2.3 Output devices*	T
	2.7 Adding text, images and multimedia elements to a presentation	P
13	2.4 How organisations use input and output devices	T
	2.8 Adding animation, transitions and additional elements	P
14	2.9 Finalising your presentation	P
	Consolidate and review all Unit 2 sessions	T&P
15	Unit 2 Theory review	P
	Unit 2 Practical review	P
Unit 3		
16	*3.1 Storage devices and media*	P
	3.4 Document preparation	
17	*3.1 Storage devices and media*	T
	3.5 Adding text and images to a document (double lesson)	
18	3.2 The importance of data backup	P
	3.5 Adding text and images to a document	
19	3.3 Banking facilities today	T

		3.6 Creating a simple data model and chart	P
	20	3.7 Final document checks and presentation	P
	21	3.7 Final document checks and presentation	P
	22	Consolidate and review all Unit 3 sessions	T & P
	23	Unit 3 Theory review	T
		Unit 3 Practical review	P
Units 1 and 2			
	24	In-house revision, consolidation and testing of Units 1 and 2	T & P
Unit 4			
	25	4.1 An introduction to computer network hardware	T
		4.8 An introduction to website design	P
	26	4.2 Computer networks	T
		4.9 Using HTML to create and edit webpages	P
	27	4.3 The internet and intranets	T
		4.10 Using website authoring software [1]	P
	28	4.4 Computer networks in business environments (1)	T
		4.11 Creating stylesheets using website authoring software	P
	29	4.5 Computer networks in business environments (2)	T
		4.12 Using website authoring software [2]	P
	30	4.6 How does ICT help business communication?	T
		4.6 How does ICT help business communication?	T
	31	4.7 Keeping computer network data confidential and secure	T
		4.13 Testing and publishing a website	T
	32	Consolidate and review all Unit 4 sessions	T & P
	33	Unit 4 Theory review	T
		Unit 4 Practical review	P
Units 3 and 4			
	34,35	In-house revision, consolidation and testing of Units 3 and 4	T & P

Year 2 – covering Units 5, 6, 7 and 8 with consolidation and revision (including all Year 1 Units): 30 weeks – 60 hours

Years 1 and 2 combined hour coverage: 130

* **T**(heory)/**P**(ractical)

Week	Session	T/P*
\multicolumn{3}{c}{Unit 5}		
1	5.1 File structure and terminology	T
	5.2 Data-handling situations	T
2	5.3 Data types	T
	5.5 Creating a database	P
3	5.4 Using the database	T
	5.6 Filters, queries and creating reports	P
4	5.7 Reports and run-time calculations	P
	5.8 Complex queries and wildcards	P
5	5.9 Producing labels and business cards	P
	5.10 Summarising data for use in other software	P
6	5.11 Investigations	P
7	Consolidate and review all Unit 5 sessions	T & P
8	Unit 5 Theory review	T
	Unit 5 Practical review	P
\multicolumn{3}{c}{Unit 6}		
9	6.1 Downloading and information retrieval	T
	6.9 Data analysis: Spreadsheet modelling skills	P
10	6.2 Security, risks and reliability	T
	6.10 Developing your spreadsheet modelling skills	P
11	6.3 The web: Today and tomorrow	T
	6.11 Analysing spreadsheet data and preparing graphs	P
12	6.4 Health and safety	T
	6.11 Analysing spreadsheet data and preparing graphs	P
13	6.5 ICT and society: The effect on unemployment	T
	6.12 Importing additional data and further work with graphs	P
14	6.6 ICT and society: Commercial and ethical considerations	T
	6.12 Importing additional data and further work with graphs	P
15	6.7 ICT and society: ICT in the home	T
	6.13 Referencing an external file and searching the data	P
16	6.8 ICT and society: Modelling and simulations	T
17	Consolidate and review all Unit 6 sessions	T & P
18	Unit 6 Theory review	T
	Unit 6 Practical review	P
\multicolumn{3}{c}{Unit 7}		
20	7.1 The systems life cycle: Analysis	T
	7.2 The systems life cycle: Design and development	T
21	7.3 The systems life cycle: Testing and implementation	T
	7.4 The systems life cycle: Documentation and evaluation	T
22	7.5 Developing questionnaires and describing an existing system	P
	7.6 Designing data capture forms and reports	P
23	7.7 Developing and interpreting a model	P

24	Consolidate and review all Unit 7 sessions	T & P
25	Unit 7 Theory review	T
	Unit 7 Practical review	P
Unit 8		
26	8.1 Physical safety in an ICT-based environment	T
	8.2 esafety	T
27	8.2 esafety	T
	8.3 The security of personal and commercial data	T
28	8.4 Producing a word-processed report in response to a hypothesis	P
	8.4 Producing a word-processed report in response to a hypothesis	P
29	Consolidate and review all Unit 8 sessions	T&P
30	Unit 8 Theory review	T
	Unit 8 Practical review	P
Units 5,6,7 and 8 theory		
	In-house revision and testing of Units 5, 6, 7 and 8	

Syllabus coverage table

The following table demonstrates how this book and the Collins Cambridge IGCSE™ ICT Student's book cover the content of the Cambridge IGCSE and IGCSE (9-1) Information and Communication Technology syllabuses (0417/0983).

The left-hand column contains the topics and sub-topics that make up the syllabus. The right-hand column contains the Collins Cambridge IGCSE™ ICT session/s in which the sub-topics are covered.

1 Types and components of computer systems	
Sub-topic	Session/s
1.1 Hardware and Software	1.2, 1.3, 1.4 2.1, 2.2, 2.3 3.1 4.1 5.7
1.2 The main components of computer systems	1.3 2.1, 2.2, 2.3 3.1
1.3 Operating systems	1.4
1.4 Types of computer	1.1
1.5 Emerging technologies	1.5

2 Input and output devices	
Sub-topic	Session/s
2.1 Input devices and their uses	2.1
2.2 Direct data entry and associated devices	2.2, 2.4
2.3 Output devices and their uses	2.3, 2.4

3 Storage devices and media	
Sub-topic	Session/s
3 Storage devices and media	3.1

4 Networks and the effects of using them	
Sub-topic	Session/s
4.1 Networks	3.2 4.1, 4.2, 4.3 8.3
4.2 Network issues and communication	4.6, 4.7 8.3

5 The effects of using IT	
Sub-topic	Session/s
5.1 Microprocessor-controlled devices	5.1, 5.3, 5.7
5.2 Potential health problems related to the prolonged use of IT equipment	5.3

6 ICT Applications

Sub-topic	Session/s
6.1 Communication	1.6 4.6
6.2 Modelling Applications	5.6
6.3 Computer controlled systems	5.1
6.4 School management systems	4.4
6.5 Booking systems	4.4
6.6 Banking applications	3.3 4.5
6.7 Computers in medicine	2.3 4.5
6.8 Expert Systems	4.5
6.9 Computers in the retail industry	2.2 5.4 7.2, 7.3
6.10 Recognition systems	1.5 2.2 4.7
6.11 Satellite systems	5.2

7 The systems of life

Sub-topic	Session/s
7.1 Analysis	7.1
7.2 Design	7.2
7.3 Development and testing	7.3
7.4 Implementation	7.4
7.5 Documentation	7.5
7.6 Evaluation	7.5

8 Safety and security

Sub-topic	Session/s
8.1 Physical safety	8.1
8.2 eSafety	8.2
8.3 Security of data	8.3

9 Audience

Sub-topic	Session/s
9.1 Audience appreciation	1.6
9.2 Copyright	1.8, 1.10 5.5

10 Communication	
Sub-topic	Session/s
10.1 Communication with other ICT users using email	1.10 8.2
10.2 Effective use of the internet	4.3 8.2, 8.3

11 File management	
Sub-topic	Session/s
11.1 Manage files effectively	1.7 2.5 3.7 4.8, 4.12
11.2 Reduce file sizes for storage or transmission	2.9 3.7

12 Images	
Sub-topic	Session/s
12 Images	1.9 2.6, 2.7 4.12

13 Layout	
Sub-topic	Session/s
13.1 Create or edit a document	1.7, 1.9 3.5
13.2 Tables	1.9 3.5 5.8
13.3 Headers and footers	1.7 3.4

14 Styles	
Sub-topic	Session/s
14 Styles	1.6, 1.7 2.6 3.5

15 Proofing	
Sub-topic	Session/s
15.1 Software tools	3.7 5.8 6.5 7.2
15.2 Proofing techniques	3.7 7.2, 7.3

16 Graphics and charts	
Sub-topic	Session/s
16 Graphics and charts	3.6 5.10, 5.11

17 Document production	
Sub-topic	Session/s
17 Document production	1.7 2.6 3.4, 3.5 4.12

18 Databases	
Sub-topic	Session/s
18.1 Create a database structure	1.7 6.1, 6.2, 6.3, 6.5, 6.6, 6.7
18.2 Manipulate date	6.4, 6.8, 6.9, 6.10
18.3 Present Data	6.8, 6.9, 6.11

19 Presentations	
Sub-topic	Session/s
19 Presentations	2.5, 2.6, 2.7, 2.8 2.9

20 Spreadsheets	
Sub-topic	Session/s
20.1 Create a data model	3.6 5.8, 5.9, 5.10, 5.11 5.12 7.7
20.2 Manipulate data	5.11, 5.12
20.3 Present Data	3.6, 3.7 5.8, 5.9, 5.10 5.12

21 Website authoring	
Sub-topic	Session/s
21.1 Web development layers	4.8
21.2 Create a web page	4.8, 4.9, 4.11, 4.12
21.3 Use style sheets	4.11

LESSON PLANS

1.1 Types of computers

This session describes the more common types of computer and introduces the theme of ICT development that features throughout the Student's Book.

Learning aims
- Identify different types of computer.
- Describe recent developments in ICT.

Differentiated learning outcomes

- **All students must** use the key words highlighted in the Student's Book in context and recognise that there are a variety of computers defined by size that are in common use.
- **Most students should** confidently use the terminology in the session; state the differences between, and the uses of, the different types of computer; identify the impact of new technology.
- **Some students could** confidently use the terminology in the session to describe the uses, advantages and disadvantages of each type of computer; and could explore the impact that advances in new technology is making.

Resources
- **Student's Book:** pages 8–11
- **Files:**
  WS1_1a.docx
  WS1_1a_Answers.docx
  WS1_1b.docx
  WS1_1c.docx
  WS1_1c_Answers.docx

Starter suggestions

Students will have personal experience of PCs, laptops and smartphones, but may only have seen mainframes or supercomputers depicted in films or on TV. Open a conversation about the processing needs of a family compared with a supermarket. Aspects to cover include:

- the amount of information that needs to be stored
- speed of processing
- need for multitasking
- need for simultaneous access by several users.

This will help students to appreciate the need for a variety of computer power.

Main lesson activities

Student task: Ask students to start creating an ICT dictionary with the highlighted key terms and a short definition for each. They can keep this up throughout the course, especially in the theory sessions. Ask them to complete **WS1_1a** and write the terms, together with the description, into their dictionary. (The solution to the puzzle is provided see **WS1_1a_Answers**.)

Discussion: As you work through the section on types of computer (Student's Book pages 8–11), discuss examples of where a large powerful mainframe computer or server might be used rather than lots of individual computers.

Discussion: Find out how many students have used smartphone and then list the things that they regularly do with them. Why is it called a smartphone?

Discussion: List the devices that students have used with a touchscreen. Discuss what features the touchscreen enables. How would those functions have been done previously (if at all)?

Student task: Ask students to carry out **Activity 1** on page 11, using **WS1_1b**. Tell students to consider tablets or phablet, alongside laptops. If the choice was between a desktop and a tablet, would their answer be different?

Give extra support by displaying pictures of mainframe computers in use so that students have a visual aid as well as the more technical information.

Give extra challenge by asking students to find users of supercomputers. What was/is the Deep Blue computer? What is the World Community Grid? Volunteer computing?

Plenary suggestions

Hold one (or more) of the following sessions:

- Ask students to identify the computer from the statements – for example: What type of computer has standardised, inexpensive spare parts?
- Student task: Ask students to complete **WS1_1c**. (The answers are provided on the Teacher Guide digital download – see **WS1_1c_Answers**.)

1.2 Hardware and software

This session will help students name peripheral devices (hardware) and be able to make a distinction between different types of program (software).

Learning aims
- Define hardware and software, giving examples of each.
- Describe the difference between hardware and software.

Differentiated learning outcomes
- **All students must** use the key words highlighted in the Student's Book in context; be able to state the difference between hardware and software; be able to identify different peripheral devices; be able to describe application and systems software types.
- **Most students should** confidently use the terminology in the session; describe the difference between hardware and software; identify internal and external hardware components; distinguish between different software types.
- **Some students could** confidently use the terminology in the session to describe different peripheral devices; identify and describe the use of internal and external hardware components; identify and describe different software types and uses.

Resources
- **Student's Book:** pages 12–13
- **Files:**
  WS1_2a.docx
  WS1_2a_Answers.docx
  WS1_2b.docx
  WS1_2b_Answers.docx
  WS1_2c.docx
  WS1_2c_Answers.docx
  WS1_2d.docx
  PPT1_2a.pptx

Starter suggestions

Students will be familiar with particular peripheral devices (mouse, printer, joystick), so have as many unfamiliar ones available as you can (such as a barcode reader and tracker ball). Try and get hold of the boxes for a variety of software: *Office*, anti-virus, anti-spyware and graphics, as well as *Windows*. Students will probably be unfamiliar with internal hardware, so have a good selection of pictures available. Ideally see if the IT team will let you have an old computer tower that can be explored.

- How many examples of hardware can students name?
- Can they provide real examples of application types?
- Why is Windows an example of system software?

Main lesson activities

This is a very simple and self-contained lesson that requires students to understand how to distinguish been the main categories of hardware vs software, and applications vs system software.

Student task: Ask students to carry out **Activity 1** using **WS1_2a**. (The answers are provided on the Teacher Guide digital download – see **WS1_2a_Answers**.)

Discussion: Use a diagram such as the one shown (use **PPT1_2a**) as a starting point. Leave space in the boxes to write a one-sentence definition.

Mind map: What devices can students name to create a list under the two types of 'Hardware' box? Can they invent a question that allows them to decide whether some software is applications or system? Form lists under the two 'Software' boxes.

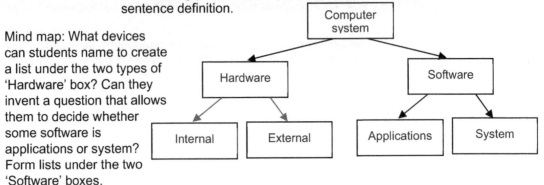

The difference between hardware and software is straightforward: if you can touch it, it's hardware; if you cannot, it is software.

Student task: Ask students to complete **WS1_2b** and discuss choices with another student.

Discussion: Do all students agree with choices made for **WS1_2b**?

Student task: Ask students to carry out **Activity 2** using **WS1_2c**.

Discussion: Do all students agree with choices made for **WS1_2c**?

Remember many examples can be confusing to students, for example a CD, DVD and Blu-ray disc look very similar. A disc containing word processing software contains software which is itself a physical object, therefore hardware.

Student task: Working in pairs, students should complete **WS1_2d**. Ask them to list all the different applications software available on the computer they are using, identifying the type of software it is and its name. You could display a completed sheet to the class towards the end of the session so that students can check their findings.

Give extra support by having pictures of all the devices that you will introduce in Unit 2 (Input and output devices) and Unit 3 (Storage devices and media) available so that if suggestions for the peripheral list start drying up, you can prompt a response.

Give extra challenge by asking students to make a list of all the hardware devices and components that they have identified in **WS1_2a and 2b** and those from the pictures of devices in Unit 2 and Unit 3 that you have available. Alongside each device ask students to state whether it is an input device, an output device or one that fulfils both functions.

Plenary suggestions

Round class quiz: You start by naming a peripheral. The first student to identify it correctly as hardware then names either another peripheral or an item of software. The next student who correctly identifies this as hardware, applications software or system software then gets to name another item, and so on.

Have a variety of pictures and labels on an interactive whiteboard and students take turns to drag images to correct labels.

1.3 The main components of a computer

This session introduces the processor as the central part of the computer system and identifies and explains the different memory and storage features of the processor.

Learning aims
- Identify the main components of a general-purpose computer.

Differentiated learning outcomes
- **All students must** use the key words highlighted in the Student's Book in context; they must be able to identify the main components of a computer system and distinguish between input and output devices.
- **Most students should** confidently use the terminology in the session; describe the uses of the components of a computer system, making a distinction between memory and storage, input and output.
- **Some students could** confidently use the terminology in the session to describe the different components of a computer system making a distinction between – and being able to compare – memory and storage, input and output devices.

Resources
- **Student's Book:** pages 14–16
- **Files:**
  PPT1_3a.pptx
  WS1_3a.docx
  WS1_3b.docx
  WS1_3b_Answers.docx

Starter suggestions

Find out what students understand about storage of data. Invariably they will talk about hard disc drives, flash cards, memory sticks and the like. Explore what they think RAM is – it is a key part of advertising a desktop or laptop – is it that RAM is just something it is better to have a lot of, irrespective of what it does? How do they think a computer knows what to do when it is switched on? Where are the instructions?

Main lesson activities

Student task: Using the diagram on page 14 (and **PPT1_3a**), students should list three peripheral devices that are examples for each of input, output and backing store devices.

Activity 1 on page 15 could need a lot of discussion. For example:

- A dishwasher is clearly a labour-saving device because it is carrying out the task of washing up; hence a manual task is now automated.
- An oven is also a labour-saving (or, perhaps more precisely, timesaving) device because, for example, we can set timers and alarms. We can set the oven to switch itself on and perform certain actions so that when we arrive home the meal is cooked. Some ovens can automatically turn food on a rotisserie saving us the task of visiting the oven periodically to turn the food manually.
- TV does not automate anything we used to do; it created a new activity.

Discussion: After completing **Activity 1**, highlight that the other devices are controlled by microprocessors as well. Ask students why they have decided that they are not labour-saving?

Describe the sequence of starting up a computer, finding devices, loading the OS and the subsequent loading and using an application. Use students to represent BIOS, OS, RAM, a disc drive and a word processing application, and appoint someone to act as a control signal fetching the OS from HDD to RAM, the application from HDD to RAM, and so on. Have one student act as the user, issuing instructions on paper to the control signal student.

You could try a data-flow diagram, but this could be confusing as the element of time has to be introduced to an event-driven cycle.

Student task: Ask students to complete **Activity 3** using **WS1_3a**.

Give extra support by having a large sheet of paper that represents RAM. Illustrate, by drawing areas on the sheet for the OS, applications and user data, the loading of program or data into RAM and its removal when the application is closed.

Give extra challenge by asking students to find out what type of memory chip the BIOS of a computer is saved on. Answer: Flash memory. Is it the kind they expected? Why is this type of memory used when it is supposed to be read only memory? Answer: Allows upgrades to be easily carried out.

Plenary suggestions

- Hold a Q&A session: show pictures of input/output devices and ask students to name them.
- Ask students how many external storage devices they can name.
- Encourage students to update their ICT dictionary.
- Ask students to complete **WS1_3b**. (The answers are provided on the Teacher Guide digital download – see **WS1_3b_Answers**.)

1.4 Operating systems

This session outlines what an OS does and the two most common methods of interacting with it.

Learning aims
- Identify different types of operating systems: CLI and GUI.

Differentiated learning outcomes
- **All students must** use the highlighted key words in context; be able to distinguish a CLI from a GUI; be able to state the function of an operating system.
- **Most students should** confidently use the terminology in the session, describe the on-screen layout of a CLI and a GUI, and explain the purpose of an operating system.
- **Some students could** confidently use the terminology in the session to describe how a CLI differs from a GUI, offering advantages of each type, and could explain the purpose of an operating system.

Resources
- **Student's Book:** pages 17–19
- **Files:**
  PPT1_4a.pptx
  WS1_4a.docx
  WS1_4a.docx_Answers

Starter suggestions

Ask students to reflect on the previous session and list the tasks that they think are necessary for an application program to be loaded into RAM and work for them.

- What happens when there is an error?
- What happens when we issue a save command?
- What happens when we want to print?

All these tasks are controlled by the operating system.

Main lesson activities

Making students familiar with the layout and terminology of a GUI should not prove difficult as terms such as desktop, icons, window, and shortcuts are in such common use.

Student task: Ask students to create folders, create new documents, delete them and trash them. Ask for comments on how easy these things are to do, and how little they need to know about how or where the computer stores folders and files.

Demonstration with Q&A: Demonstrate the difference between clicking on an icon to show the contents of a folder and doing the same through the **Command Prompt** (or your equivalent CLI).

- With DOS type in **dir x:** (where x is the letter of the drive to list the contents)
- Creating a new folder? **mkdir f:\iCTGCSE**
- Emphasise to students that they have to use the syntax precisely, otherwise the command will simply return an error message. There is no visual aid or suggested help.

Session 1.1 introduced the different types of computer or server, so students should be comfortable with the idea that there will be different OSs to make these computers do the variety of complex tasks they were built for.

Discussion: Discuss the tasks that a mainframe computer or server, may need to handle. Mention multitasking and multiprocessing, using a bank or supermarket as an example.

Student task: Ask students to complete **Activity 1** by filling in **WS1_4a**.

Discussion: Both CLI and GUI are screen and keyboard based. Operating systems now include gesture and dialogue-based systems. How many devices in students' homes can now be operated using touchscreens (tablets and smartphones) or via voice commands (smart speakers for example)?

Student task: Ask students to complete **Activity 2**.

Give extra support by using the diagram below (PPT1_4a), which indicates that the OS is protecting the hardware, but also acting as a bridge between the hardware and third-party software. Ask students to think of tasks that might need to pass a signal or data from one layer to another. What about saving data? Printing a report? Printer out of ink?

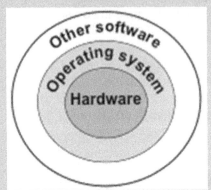

Give extra challenge by asking students to consider the advantages and disadvantages of direct keyboard and mouse computer access compared to touch and dialogue. Answers might include the lack of advanced functionality or ease of use.

Plenary suggestions

Run through student answers to **Activity 1** (Student's Book, page 19). Make a larger grid where you can collate all their different answers. Find out which desktop OSs and which smartphone OSs they have.

1.5 The impact of emerging technologies

This session identifies some of the technology that is starting to, or soon will, impact on our everyday lives. As time and technology move forward, new breakthroughs start to capture the imagination, and soon ideas become reality.

Learning aims
- Describe how a range of emerging technologies are being introduced into our everyday life, and the impact this is having.
- Consideration of technologies that are breaking through and how they could be employed in our lives.

Differentiated learning outcomes
- **All students must** be able to identify new technology and describe where it is being used.
- **Most students should** be able to identify where new technology has been introduced, assessing its usefulness; identify and consider the newly emerging technologies.
- **Some students could** discuss the application of new technologies in our everyday lives; identify and discus the impact the newly emerging technologies could have on our lives.

Resources
- **Student's Book:** pages 20–22

Starter suggestions

Begin with a discussion about artificial intelligence and robotics.
- Do students know what these are?
- What kind of a future developments might there be in these fields?

Lead a discussion about new technology including new hardware 'techniques' as well as new software for new applications.

Main lesson activities

This session is heavy on internet research and vocabulary. It might be useful for students to have looked at sections of Session 1.8 (Internet research) beforehand.

Read through the section headed **Security**.

Discussion: have students seen any application of **artificial intelligence biometrics**? Airport security is increasingly becoming reliant on this technology, with iris scanning becoming more widely used. Some cars already have fingerprint recognition in the door handles. Which?

Activities 1 and **2** can be done.

For **Activity 2**, depending on the student group, you might want to have a class discussion focussing on one area, or let students work in pairs on an area. Student pairs could then produce a report on that area to feed back to the class.

Teacher task: Text translation may be of particular interest in international school settings, so before this session it might be useful to investigate how your school manages the translation task of, for example the prospectus, or website content. Do not be disappointed to find that it is done completely manually. Instead, you could use this as an opportunity to explore the translation possibilities for your school.

Teacher task: prepare a small paragraph of text to be submitted to an online translation tool, or smartphone application.

Discussion: How is a school prospectus or website content translated? What might need to be done if the text to translate is particularly long? Have students got a grasp of how computer–assisted translation is done? Also show examples of websites where the language shown is automatically translated. Show **www.apple.com** for example, the language can be changed at the bottom of the home page.

Students can complete **Activities 3 and 4**, feeding back to the class.

> **Give extra support** by getting students into small groups and then brainstorming search terms to use in Activities 1 and 2. From search results work first on creating a definition and then listing examples of where these technologies can be found.
>
> **Give extra challenge** by asking students to work in pairs to produce a poster advertising a car with the latest emerging technology features. Place suitable textboxes alongside parts of the car where the technology is located explaining the advantages of that technology.

Plenary suggestions

This session has been about exploration and discovery as much as anything else. There have been a number of feedback opportunities in this session, so try and draw out any conclusions or summarise what has been researched and discovered.

1.6 Using a computer system to communicate *(double lesson)*

Learning aims
- Understand communication applications used to create newsletters and posters, websites, multimedia presentations, audio, video, media streaming, e-publications.

Differentiated learning outcomes
- **All students must** be able to use the key words highlighted in the Student's Book in context and state the purpose of the seven most common forms of communication.
- **Most students should** be able to use the terminology in the session with confidence, describe situations where each of the seven most commonly used forms of communication might be used and be able to describe what corporate style refers to.
- **Some students could** confidently use the terminology in the session and evaluate which would be the most appropriate method of communication in a given situation and be able to discuss the importance of corporate style.

Resources
- **Student's Book:** pages 23–28

Starter suggestions

Explain that many factors play a part in deciding the most appropriate method of communication, including the purpose, the information to be given, and the target audience.

Have available examples of newsletters and posters that advertise a range of events, information and places. Discuss their suitability for that purpose based on context – the information being given, and the audience being targeted. Why do students think that these are appropriate formats? Can they come up with some guidelines for choosing the right format?

Main lesson activities

There are a number of varied tasks within this session, but with good examples it will be possible to illustrate the advantages or suitability of a method without too much direction.

Newsletters and posters

Build on the starter activity by asking students to complete **Activity 1** (Student's Book, page 23). Come to some agreement about content. Have they all identified the necessary detail? (Compare to some real-life online examples)

Student task: The restaurant needs a poster to advertise their opening night. What does the poster need to contain? How can it be made to catch people's attention?

Student discussion in pairs – feedback to whole class: After the restaurant has been open for a few months, the owners decide to expand their range of promotional material.

- What other documents could they create?
- What content could these documents contain?

- How might these documents be distributed?

Corporate style and software choices

Student task: **Activity 2**. Discuss the results of this activity and compare them to the examples discussed in the Student's Book. It is important to discuss the cost of professional software packages and how this can be prohibitive to many people. If access to any of the example software isn't available, most will have promotion videos on their websites that give examples of functionality.

Student task: **Activity 3**. Break down the corporate style of the Student's Book and discuss how a similar document could be created using software available in school.

> **Give extra support** by asking students to create a mock-up of the restaurant poster. Together with the presentation 'story board', this gives students three draft examples. This is reinforced in this unit's practical sessions and in the practical parts of Units 2 and 4.
>
> **Give extra challenge** by asking students to draw up a design for the homepage of the restaurant's website. How often would the page be updated? What could be included that gives viewers confidence in the page content?

Plenary suggestions

Ask all students to comment on the designs for the presentation and the webpage by getting students to present their ideas to the class. As a class, discuss the strengths and weaknesses of the designs.

1.6 cont. Using a computer system to communicate (double lesson)

Learning aims
- newsletters and posters, websites, multimedia presentations, audio, video, media streaming, e-publications.

Differentiated learning outcomes
- **All students must** be able to use the key words highlighted in the Student's Book in context and state the purpose of the seven most common forms of communication.
- **Most students should** be able to use the terminology in the session with confidence, describe situations where each of the seven most commonly used forms of communication might be used and be able to describe what corporate style refers to.
- **Some students could** confidently use the terminology in the session, evaluate which would be the most appropriate method of communication in a given situation and be able to discuss the importance of corporate style.

Resources
- **Student's Book:** pages 23–28
- **Files:**
  WS1_6a
  WS1_6b

Starter suggestions

On the whiteboard, create a mind-map with the word 'Communication' at the centre and add around it 'Posters' and 'Newsletters'. Ask for as many additional examples of how we can communicate with others using documents. Compare the answers to the list covered in this lesson.

Main lesson activities

This half of the double lesson focuses on websites, multimedia presentations, audio, video, media streaming and e-publications.

Websites

After students have completed **Activity 4** (Student's Book, page 25), ask them to discuss their findings in pairs or small groups:

- Are they surprised by what they have discovered?
- Have they found sites that have completely out-of-date information?
- What does this tell them about the website and its owners?
- What are their favourite websites? What is it about these websites that they like?
- What kind of content do they have?
- What mix of text/images/animation do they have?
- What makes these websites attractive?

Multimedia

Small group activity:

- Revisit the restaurant idea from the first part of this session. Another idea the owners have is to put up a screen in the restaurant waiting area that will display a continuously looping presentation, advertising special events and offers as well as the food that is on the menu:

    Is this a good idea? What are the advantages and disadvantages?

    Should the presentation be mainly text?

    Should it have images? What about sound? Or animation?

- Draw up a storyboard of the slides' content.

After completing this task, each group should present their ideas to the rest of the class.

Student task: Ask students to work through **Activity 5** on page 26, comparing and contrasting two types of document in the scenario given. Students could work in pairs. Feed back to the whole group; ask each pair in turn to suggest one contribution.

Audio, video and media streaming

Teacher demonstration may be needed here:

- If your school has an audio/visual department, try and demonstrate some of the work they do and the applications they use.
- Show examples of current media streaming services for music, video, podcasts. How are organisations using these platforms for promotion?

Student task: Ask students to work through **Activity 6** on page 28, using **WS1_6a** to note down their answers and their reasons. Then, as a whole group, work through the scenarios taking a show of hands as to which suggestion is best for each scenario. When there is general agreement about which option is best, discuss the reasons why. If time allows, create additional scenarios for media-streaming and e-publications.

Student task: Ask students to complete **WS1_6b**, working through the tasks in small groups. Depending on the time you have available, you could set different tasks for different groups and then ask each group to feed back to the whole class. If you are very short of time, this could be split and used as a homework activity.

Plenary suggestions

Discussion: Ask students to consider events, such as a school concert or a parents' evening. How would these events be best advertised? When is it appropriate to choose a printed paper-based document over an electronic document and vice-versa?

1.7 Document production: Entering and editing data from different sources

This session demonstrates how to locate and import text that is in other formats before working with page, paragraph and character settings.

Learning aims

- Create and open documents from different sources.
- Manipulate the page layout: size, orientation, headers and footers.
- Work with paragraph and character settings: alignment, font style and emphasis.

Differentiated learning outcomes

- **All students must** be able to import text from another source and adjust paragraph and character settings.
- **Most students should** be able to create a document from text in any format and confidently adjust page, paragraph and character settings.
- **Some students could** locate and import text from any source, creating a document exactly to given specifications.

Resources

- **Student's Book:** pages 31–36
- **Files:**
  WS1_7a.docx
  WS1_7a_Answers.docx
  ComputerTypes.rtf
  DesktopComputer.txt

Starter suggestions

Begin with a discussion about file extensions.

- Do students know what these are?
- Do they know what extensions their own software uses for word processed documents?
- What other extensions do they know? (.jpg and .mp3 probably)

Lead into asking how information could be shared between different people if they do not use the same software. Introduce the idea of generic file types.

Main lesson activities

Teacher/systems preparation: You may need to alter the settings of the students' computers so that you can see the file extensions when the contents of a folder are viewed. (Look for a setting under **Folder options** where you can either **Hide** or **Unhide** extensions of known file types.)

Student task: Ask students to open their own work area and identify various extensions (**Activity 1**, Student's Book, page 31) while you compile a list on the board. Can students identify applications that use these extensions?

Student task: Ask the students to follow the steps in the Student's Book, pages 31–32 for opening the document **ComputerTypes.rtf,** and then complete **Activity 2**. They should not worry if the text of **ComputerTypes.rtf** does not fit on one page; the important part of this activity is being able to find and understand the settings. They can then adjust the settings in **Activity 3**.

Student task: Ask students to complete **Activity 4** (Student's Book, page 33) using **WS1_7a**. Knowing and being able to identify serif and sans-serif fonts is an important skill.

Teacher supervision: Make sure you can direct students to open the header and footer via the menus and tabs because sometimes they are awkward to open up with a double click.

Teacher supervision: Emphasise the need to save work regularly. A good habit might be after entering a new paragraph or making a significant change. Depending on the version of software being used, it might also be worth demonstrating the Auto-Save function.

> **Give extra support** by having a copy of **ComputerTypes** with **DesktopComputer.txt** already inserted. This will enable students to practise fully the editing and formatting tasks, should the import be difficult. Ask students to make other, specific changes to the format of paragraphs and text. Introduce font colour.
>
> **Give extra challenge** by asking students to work in pairs. One student asks the other to make a specific change to their text. They then swap roles, exploring the other buttons in the paragraph and font panels.

Plenary suggestions

After saving their work at the end of **Activity 7** (Student's Book, page 35), students should then also resave the document as: **ComputerTypes.txt**

Ask students to open this and see the effect that saving in this format has had. A basic (txt) text file contains only very basic formatting, much of their formatting in this lesson will be removed.

1.8 Internet research

This session shows students how to find information, how to download and save relevant information, and how to place that information into a document.

Learning aims
- Locate information from a given website URL.
- Find specified information using a search engine.
- Download and save information.
- Import and place information from a variety of external sources in a document.

Differentiated learning outcomes
- **All students must be able to** locate a website from a given URL; use a search engine to find information; download and save information from a website.
- **Most students should** locate a website from a given URL; use a search engine's advanced features to find information; download, save and import information to a document.
- **Some students could** find relevant and accurate information from either a given URL or using logical searching techniques; download and save and import information to a document.

Resources
- **Student's Book:** pages 37–42
- **Files:**
  WS1_8a.docx

Starter suggestions

Begin with a discussion about search engines.
- Do students know what they are?
- Do students know how, in principle, they work?
- Ask for examples of searches that have turned up results that were not what was expected, or which returned all sorts of unrelated items.

Main lesson activities

Teacher advice: Use Google as the demonstration tool throughout this session so that what students can see will largely equate to the images and commands in the Student's Book. It is important to remind students that there are other search engines.

Once you have demonstrated and students have completed **Activity 2** (Student's Book, page 38), ask them to try the same search but using another search engine and compare results. Explore the idea of indexing and the algorithms that are used – point out that each company has different algorithms and criteria for their searches.

Student task: For **Activity 3** (Student's Book, page 38), show students how to access the 'Advanced Search' options of different search engines. Although useful for specific searches, these are now replaced by algorithms integrated with users search history.

Q&A session: Do the advanced features of the different search engines use the same refining tools?

Teacher advice: Boolean operators (**Activity 4**) will be encountered again when using databases (Unit **6**), so it is good to ensure that students are clear about them at this early stage.

Ensure that students have created and are using the folder NEWSLETTER and the **NewsletterURLs** document, as these are referred to a number of times – and not just in this session.

Emphasise the issue of copyright. Students can download text as material to use in an edited, changed form – not just a straight copy and paste. See Session 5.5 for more information about copyright law.

Introduce the idea of hot keys – keyboard shortcuts that save you finding and clicking on icons: Ctrl+C, Ctrl+V for **Copy** and **Paste**. (Emphasise V for **Paste** and not P, which opens the printer dialogue box.) Encourage students to start creating a list of these in their ICT dictionary.

Student task: Ask students to work through the tasks in **WS1_8a**.

> **Give extra challenge** by asking students to search just the Wikipedia website for the five devices. Are the results useful? Get them to search Google for images of the five devices, starting with the key term and adding refining terms, as they did with the normal web search. Do results get better? Can they suggest a reason for the difference in relevance?

Plenary suggestions

Ask students to compare the advanced search keywords and criteria they used to find each of the text segments that they have saved in **NewsletterURLs**.

- Which search seems to have produced the most useful text?
- Do students see any pattern in the way that search criteria or keywords are best used?

Ask students to compare the contents of their **NewsletterURLs** document and discuss why they have chosen particular pieces of text.

1.9 Creating a newsletter

This session takes students through the various stages involved in mastering and applying the techniques needed to create attractive-looking documents.

Learning aims
- Create and open documents with information from different sources.
- Import, place and manipulate a variety of information from a variety of sources.
- Format the page layout: size, orientation, margins, columns, headers, footers, text alignment.

Differentiated learning outcomes
- **All students must be able to** create and open documents with information from different sources; change page size, orientation, header and footers; import images.
- **Most students should** create and open documents with information from different sources; change all aspects of a page layout; import images.
- **Some students could** create and open documents from any source; change all aspects of a page layout; import and place an image with text wrapping.

Resources
- **Student's Book:** pages 43–47
- **Files:**
  NewsletterText.rtf
  InkjetPrinter.jpg
  Joystick.jpg
  Keyboard.jpg
  LCDMonitor.jpg
  Mouse.jpg
  Scanner.jpg
  WS1_9a.docx

Starter suggestions

Have a look back at the document created in Session 1.8. If necessary, have copies available. Discuss what is not attractive about it, for example: no images; text format/alignment is boring. Get suggestions about how this could be improved.

Main lesson activities

Teacher advice: Ensure that you have available the image files: **Keyboard.jpg**, **Mouse.jpg**, **Scanner.jpg**, **Joystick.jpg**, **LCDMonitor.jpg**, **InkjetPrinter.jpg**, and the text document **NewsletterText.rtf** from the Student's Book digital download.

Ensure that each student has the text to work with as they start **Activity 1** (Student's Book, page 43).

Student task: Use **Activity 2** to play around with adjusting the gap between the columns as well. It doesn't matter what happens in this activity or how the text gets messed up because you have the saved version from **Activity 1**.

Teacher advice: Make sure that at the start of **Activity 3** students do have a copy matching that at the end of **Activity 1**. Have a copy on hand.

It is important that students understand the concept of cropping as opposed to squashing. Aspect ratio is vitally important and students need to understand it fully, because at a number of points in the practical tests (word processing and webpage creation), students will be asked to preserve aspect ratio after resizing.

Point out the **Tip** on page 46 of the Student's Book that mentions the very useful shortcut Ctrl+Z – ask students to note this down in their ICT dictionary.

Student task: Text wrapping can create a number of interesting results so allow students to play around to see them. The layout illustrated after **Activity 5** (Student's Book, page 47) is only a suggestion. It does not matter where images are placed, just that the document looks attractive.

Student task: Using the material they saved from **WS1_8a**, students should now create a poster as outlined in **WS1_9a**.

> **Give extra support** by taking students through the process of column creation and image manipulation step by step.
>
> **Give extra challenge** by asking students to create the newsletter in three columns so that it can be folded into thirds, as in this diagram:
>
>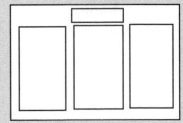
>
> Or, ask students to create the two-column version in portrait orientation.

Plenary suggestions

Ask students to list all the steps they took to transform the text that they had at the start of the session to the completed document.

- How many of these steps alter the actual layout?
- How many are about appearance only?
- What other suggestions do they have for making the document more attractive?

1.10 Communication *(double lesson)*

Learning aims
- Understand the need for, create, and apply a corporate style to a document.

Differentiated learning outcomes
- **All students must** be able to edit a document and apply a specified house style to text.
- **Most students should** edit a document, create and apply a specified house style to text.
- **Some students could** edit a document and consider, create and apply consistent house styles to document text.

Resources
- **Student's Book:** pages 48–54

Starter suggestions

Have available some examples of documents that follow a strong corporate style, these might be leaflets, websites or video advertisements. Companies like Apple, McDonalds, Disney or IKEA are strong examples to show.

If time allows, show examples with the company name missing, can students identify them through the style alone?

Main lesson activities

Discussion: refer back to **Activity 3** in Session 1.6 and the outline of corporate style; why it is important and the need for consistency in applying any styles. Ask how a document (or series of documents) could have exactly the same styles applied throughout.

Discussion: look back at the documents that you showed at the start of the lesson. What fonts are used? Are they appropriate to the audience? Are they easily read? Is there a clear difference between body text and headings? Are they consistent?

Demonstrate: create a style of the students choosing: font face, size, colour, justification, line spacing and so on.

Students should be able to carry out **Activity 1** by themselves.

> **Give extra support** to students who are having difficulty by forming them into a small group and working with them to create styles.
>
> **Give extra challenge** by asking students to work in pairs creating a new style, experimenting with addition formatting options.

Plenary suggestions

Hold a Q&A session: Display all the terms that have been introduced in this session. Ask all the students in turn to explain the terms.

Ask students to review the 'style' that they themselves normally use when preparing word processed documents. Do they use different styles depending on what they are producing? Would special care with style be needed when producing coursework for example?

1.10 cont. Communication *(double lesson)*

Learning aims
- Send and receive documents and other files electronically.
- Manage contact lists effectively.

Differentiated learning outcomes
- **All students must** be able to create, open, reply to and forward an email; attach a file to an email; download and save an attachment.
- **Most students should** create, open, reply to (using Cc and Bcc) and forward an email (with and without an attachment); add and delete contacts on a contact list.
- **Some students could** create, open, reply to (using Cc and Bcc) and forward an email (with and without an attachment); save and store files in a structured way; create and manage a contact group.

Resources
- **Student's Book:** pages 48–54

Starter suggestions

Display the terms: attachment, Cc, Bcc, reply, forward, contact list.

- How many of these do students already know?
- Can they describe the difference between **Cc** and **Bcc**?
- Can they explain the difference between **Reply** and **Forward**?

Students will have experience of emailing, but what do they actually do with their emailing software?

Main lesson activities

This session could be difficult to monitor because students may have very differing email accounts. Ideally, students have a school email account and can make use of the same software in class.

This session will be best navigated by students working from the beginning of the section **Using emails** to send attachments in small groups (of three or four) and working together step by step through the Student's Book session and completing **Activities 2**, **3** and **4**.

Ask all students to work through **Activity 5** at the same time so that all students have information to move to the next part of the session.

Demonstration: Demonstrate how to perform the task in **Activity 6** (Student's Book, page 52).

Student task: Ask students to do **Activity 6** by themselves.

Demonstration: Demonstrate **Activity 7** (Student's Book, page 52).

Student tasks: Students should now be able to do **Activity 7** by themselves. Check this and then allow them to continue on with **Activity 8**.

Once all the above tasks have been completed successfully open a discussion concerning **Emailing issues**. Are students aware of any constraints, either national (perhaps government controlled) or local (school filtering/monitoring)? How do they feel about *netiquette*? How different would an email they sent to All Computers Inc. applying for the Marketing Manager, be from an email sent to a friend telling them of a successful interview?

Activity 9 could be incorporated into the above discussion.

> **Give extra support** to students who are having difficulty by forming them into a small group and working with them through **Activities 2, 3 and 4**.
>
> **Give extra challenge** by asking students to download and store all the components used in Session 1.9 into a new subfolder called **MATERIALS** in their **NEWSLETTER** folder. Then ask them to email these components to all the people in their contact group.

Plenary suggestions

Hold a Q&A session: Display all the terms first introduced in the starter to this session. Ask all the students in turn to explain the terms.

Ask students to review their own use of email, how they express themselves in them, and whether they feel contents should be monitored. Could they ever see themselves using (or needing to use) password protection or encryption techniques?

2.1 Input devices (double lesson)

Learning aims
- Identify input devices.
- Identify suitable uses of input devices stating the advantages and disadvantages of each.

Differentiated learning outcomes
- **All students must** be able to identify a range of input devices.
- **Most students should** be able to identify a range of input devices, their uses and the relevant advantages and disadvantages of each.
- **Some students could** explain the relevant advantages and disadvantages of a full range of input devices depending on the needs of their user.

Resources
- **Student's Book:** pages 60–65
- **Files:**
  PPT2_1a.pptx
  WS2_1a.docx
  WS2_1b.docx
  WS2_1c.docx
- **Further resources:**
  Projector, selection of hardware input devices (and one output device)

Starter suggestions

Have a selection of input devices out on display at the front of the classroom, such as a microphone, mouse, keyboard and camera. Include an output device as an odd one out – for example, a speaker.

Discussion: Ask what the similarities are between the devices. Discuss how they are used, and which one does not fit into the category of input device. Write comments on the whiteboard.

Show **PPT2_1a** (Slide 1) and ask where each item fits into the diagram. Discuss analogue and digital data, and the differences between them. Point out that some of the devices in this session will require ADC converters.

Give an outline of the lesson: input devices – their uses, advantages and disadvantages.

Main lesson activities

To prepare for the first activity, use **WS2_1a** to create cards, each showing the name of an input device. Divide the class into small groups and give each group one or more cards. You can either give out all the cards or just use a selection depending on time constraints.

> The first part of the double lesson ends with this class activity.
> If this session is being delivered as two single lessons, a recap and introduction will be needed at the start of the second lesson.

Student task: Give students 20 minutes to produce a brief verbal presentation of their device (or devices) to present to the group. Their presentation should include: a description of the device, its advantages, disadvantages and possible uses. Students can gather information from the Student's Book and the internet.

Class activity: Watch as many presentations as time allows. After each one, pool together the group's comments, pointing out any errors or omissions from the Student's Book. Expand on their comments as necessary. Also prompt the class to ask questions about each presentation if appropriate.

Ask students to carry out **Activity 1** on page 64, using **WS2_1b** to note down details about five devices they have used at home and school. Afterwards discuss students' answers with the class, especially any not listed in the Student's Book.

Show **PPT2_1a** (Slide 2) and discuss the importance of using the most appropriate device for the right task. How can we decide between similar devices? Make sure students understand what the four terms in the diagram mean. There are a variety of devices that can take a photo or input data into a computer, but the best choice depends on the user and their needs. A professional photographer would not use a mobile phone camera in a photo shoot, and a bank employee entering purely numerical data may not need a full QWERTY keyboard.

Class activity: Discuss **Activity 2** on page 65 and the first example (the animated film studio). Refer to the information about the company (True to Life Studios) in the Introduction on Student's Book page 58 and discuss what devices might be used and why. Write student suggestions on the whiteboard as a mind map and keep referring to the example and 'Important considerations' diagram on page 64.

Student task: Use the class ideas to complete the first part of **Activity 2** and then complete the rest. **WS2_1c** can be used to write or type their responses.

> **Give extra support** by having students work in groups to support each other.
>
> **Give extra challenge** by asking students to explain the use of devices outside their normal experience, such as barcode readers, pet identity chip systems or industrial equipment such as CNC machines or computer-controlled climate systems.

Class discussion: Talk through each organisation, ask for student responses and discuss their findings, correcting and developing ideas as required.

Plenary suggestions

Hold a Q&A session: As a group, decide on the five devices that were thought the most difficult to understand or were discussed the least. Students could then produce a short report or presentation on these devices as homework.

Hold up the selection of devices from the beginning of the lesson and ask students to come up with some wider real-world uses for them. Question students as to why other devices would not be so suitable.

Ask students to consider how these devices could develop in the future through topics such as theme-park rides, automated transport systems or robotic manufacturing. Discuss the use of mobile phones as input devices: for example, mobile devices can use their cameras to scan QR (quick response) codes used in advertisements.

End by recapping the objectives of the lesson.

Assessment suggestions	**Activity 1** could be given as homework between the first and second parts of the lesson if the two parts are separate.

2.2 Direct data entry devices

Learning aims
- Identify direct data entry devices.
- Identify suitable uses of direct data entry devices stating the advantages and disadvantages of each.

Differentiated learning outcomes
- **All students must** be able to identify a range of direct data entry devices.
- **Most students should** be able to identify a range of direct data entry devices, their uses and the relevant advantages and disadvantages of each.
- **Some students could** explain the relevant advantages and disadvantages of a full range of direct data entry devices depending on the needs of their user.

Resources
- **Student's Book:** pages 66–68
- **Files:**
  PPT2_2a.pptx
  WS2_2a.docx
  WS2_2b.docx
- **Further resources:**
  Projector, selection of direct data entry devices if possible (and one output device)

Starter suggestions

The devices in this section may be harder to source, but the school library may have a barcode scanner. If devices can be found, then display with an output device, inviting students to try and find the odd-one out.

Discussion: Although the devices in this section are still technically input devices, they are slightly different in that they can read data directly and send the information into a system. A digital camera, for example, can take the image of someone's handwriting and simply display it, whilst an OCR reader can convert it directly into editable computer data.

Give an outline of the lesson: direct data entry devices – their uses, advantages and disadvantages.

Main lesson activities

To prepare for the first activity, use **WS2_2a** to create eight cards, each showing a direct data entry device. Divide the class into small groups and give each group one or more cards. Either use all the cards or a selection based on time constraints.

Student task: Give students 10 minutes to produce a brief verbal presentation of their device (or devices) to present to the rest of the class. Their presentation should include a description of the device, its advantages, disadvantages and possible uses. Students can use the Student's Book and the internet to collect information.

Class activity: Watch as many presentations as time allows. After each one, pull together the groups' comments, pointing out any errors or omissions from the Student's Book. Expand on their comments as necessary. Also prompt the class to ask questions about each presentation if appropriate.

Ask students to carry out **Activity 1** on page 67, using **WS2_2b** to note down experiences they have had with any of the devices and discuss some of their answers.

Discussion: Show **PPT2_2a** and discuss the importance of using the most appropriate device for the right task. It is important to stress that these devices often deal with sensitive and important data such as exam papers, bankcards, cheques and product barcodes. What are the implications if information is incorrectly read using these devices?

The first part of the double lesson ends with this class activity. If this session is being delivered as two single lessons, a recap and introduction will be needed at the start of the second lesson.	Student task: Following the discussion, ask students to complete **Activity 2** on page 68. **WS2_2b** can be used to note their ideas. Additional guidance may be required, as students may not have experience personal use of the devices in this session. Class discussion: Talk through some of the responses, correcting and developing their ideas as required.

Give extra support by having students work in groups to support each other.

Give extra challenge by asking students to look specifically at the technology behind and key differences between Chip and PIN and Near Field Communication systems for example.

Plenary suggestions

Hold a Q&A session: As a group decide on the three devices that were thought the most difficult to understand or were discussed the least. Students could then produce a short report or presentation on these devices as homework.

Ask students to consider the impact these devices have made on our society over the last few years, from shopping to banking and how might they continue to develop in the future. RFID tags for example have raised security concerns as they provide a method of tracking both objects and people.

End by recapping the objectives of the lesson.

Assessment suggestions	Homework could be given on the use RFID tags and how they might revolutionise the tracking of both products, and people, all around the world.

2.3 Output devices *(double lesson)*

Learning aims
- Identify output devices.
- Identify suitable uses of the output devices stating the advantages and disadvantages of each.

Differentiated learning outcomes
- **All students must** be able to identify a range of common output devices, such as displays and printers, from a given variety of input and output devices.
- **Most students should** be able to identify a range of output devices, their uses and the relevant advantages and disadvantages of each.
- **Some students could** explain the relevant advantages and disadvantages of the complete range of output devices outlined in the Student's Book, depending on the needs of their user.

Resources
- **Student's Book:** pages 69–73
- **Files:**
  PPT2_3a.pptx
  WS2_3a.docx
  WS2_3b.docx
  WS2_3c.docx
  WS2_3d_Answers.docx
  WS2_3d.docx
- **Further resources:** Projector, stickers/sticky notes

Starter suggestions

Before the lesson, place a sticker or sticky note on all the output devices in the classroom; monitors, speakers, printers, bells, buzzers, etc. Ask students to retrieve them one at a time (or all at once in a game). Once all have been found, students must explain what each one is and why they think it is an output device.

Show **PPT2_3a** (Slide 1) of a system and discuss the part these devices play in the system. Give an outline of the lesson: output devices – their uses, advantages and disadvantages.

Main lesson activities

In a similar way to lesson 1, use **WS2_3a** to create cards, each showing the name of an output device. Divide the class into small groups of two or three and give each group a different output device.

> The first part of the double lesson ends with this class activity.
>
> If this session is being delivered as two single lessons, a recap and introduction will be needed at the start of the second lesson.

Student task: Students have 20 minutes to produce a brief presentation of their device that includes: a description of the device, its advantages, disadvantages and possible uses. (The presentation can be verbal or electronic if time allows, for example, using PowerPoint.) Students can use the Student's Book and internet if available. Depending on the group, a choice could be offered to provide straight information on the devices or to present more detailed information based on a scenario such as a vets' practice, supermarket or garage.

Class activity: Watch as many presentations as time allows. After each one, pull together the group's comments, pointing out any errors or omissions from the Student's Book. Expand on their comments as necessary.

If a plenary is required, discuss what students thought were the most difficult concepts to comprehend and note these as a focus for the second lesson.

Ask students to carry out **Activity 2** on page 73, using **WS2_3b** to note down details about five devices they have used at home and school. Afterwards, discuss students' answers with the class, especially any not listed in the Student's Book.

Show **PPT2_3a** (Slide 2). Discuss how, when choosing devices, each one should be considered in terms of speed, relevance, quality and cost. Make sure students understand what these four terms mean. Different devices can perform similar tasks, but the needs of the user should be considered to decide on the most practical choice.

Student task: Ask students to carry out **Activity 3** on page 73, using **WS2_3c** to evaluate the three visual devices and the four printing devices. Each device should be compared to the others in the group using the terms discussed. Information can be sourced from the earlier presentations, the internet or the Student's Book.

Class activity: Discuss **Activity 4** on page 73 and spend a few minutes on the first bullet point (the animated film studio), discussing what devices might be used in that organisation and why. Write students' suggestions on the whiteboard as a mind map and keep referring to the 'Considerations' diagram on **PPT2_3a** (Slide 2).

Student task: Ask students to use the class ideas to complete the first part of **Activity 4** using **WS_2_3d** to write or type their responses. They should then complete the Worksheet for the other organisations listed.

Peer assessment: Ask students to swap their responses to **WS_2_3d** and assess each other's responses. Let students see the sample answers given on **WS_2_3d_Answers**. Answers provided outside the ones given in the Student's Book should be discussed as a group rather than discounted.

> Give extra support by choosing group members who can support each other.
>
> Give extra challenge by asking students to explain the use of devices outside their normal experience.

Plenary suggestions

Hold a Q&A session: As a group decide on the five devices that were thought the most difficult to understand or were discussed the least. Students could then produce a short report or presentation on these devices as homework.

Discussion: Ask students to read **Activity 1** and think of as many examples as they can of the use of robots and similar devices in dangerous situations. This could also be linked into the theme of future development.

Discussion ideas: What are the output devices of the future? How might these devices change in the future? What could follow on from 3D cinema and television? Whatever happened to virtual reality headsets and holograms?

Assessment suggestions	Activity 1 could be given as homework between the first and second parts of the lesson if the two parts are separate.

2.4 How organisations use input, direct data and output devices

Learning aims
- Understand ICT in everyday life, in terms of data handling and control applications.
- Understand batch, online and real-time applications used in school management and booking systems, libraries and retail.
- Identify suitable uses of input and output devices stating the advantages and disadvantages of each.

Differentiated learning outcomes
- **All students must** be able to identify a range of organisation-based situations and some of the devices used within them.
- **Most students should** be able to describe how different devices are used in ICT systems and identify the information processed within them.
- **Some students could** explain how information is input, processed and output within a system, and also consider the most appropriate devices based on user requirements.

Resources
- **Student's Book:** pages 74–75
- **Files:**
  PPT2_4a.pptx
  PPT2_4b.pptx
- **Further resources:** Projector

Starter suggestions

Role-play activity: Divide the class into three groups and ask them to prepare and deliver a short scene, based on the points below, where an experienced employee is explaining how these systems work to new workers:

- setting and tripping a burglar alarm
- a medical research centre
- shopping using a credit or debit card
- an electronic school registration system.

Students may well make lots of mistakes, but this activity can be discussed in the plenary to test new learning. Students should try and mention as many input and output devices previously discussed as possible.

Student can refer to the Student's Book and internet as required.

If this role play is not suitable, an alternative could be to sketch a storyboard or comic book of the scene.

Give an outline of the lesson: how organisations use input and output devices in their systems.

Main lesson activities

Student task: Ask students to carry out **Activity 1** on Student's Book page 75. They should write down lists of devices associated with the three scenarios. Their answers could be either typed or handwritten. Discuss students' answers with the class.

Show and discuss **PPT2_4a** (Slides 1 and 2): Concentrate on the processing taking place in the two systems, and how information is analysed and output in different forms. The hardware possibilities constantly change as new devices are developed. Can students suggest any devices not listed in the display?

Demonstration: Take one of the scenarios in **Activity 2** and create a diagram on the whiteboard similar to that on page 75 of the Student's Book. Ask students to contribute verbally or to add parts to the diagram themselves. Remind students that this activity is based around the unit scenario of an animated film studio (see Student's Book page 58 for background information).

Student task: Ask students to complete **Activity 2**, creating diagrams for the other scenarios, either on paper or electronically. If possible, project student examples to the class or hold them up to discuss. Versions of all four diagrams are shown in **PPT2_4b**.

Give extra support by explaining difficult keywords and concepts: microprocessor, actuator, ISBN, sensor-based systems.

Give extra challenge by asking students to identify and describe additional systems outside their home and school life, or you could offer additional suggestions, such as central-heating systems, voting or university application systems.

Plenary suggestions

How would the role-play activity at the beginning of the lesson be different now? What sort of extra detail could students now add?

Look again at the three scenarios (burglar alarm, shopping and school registration) at the beginning of the session. What happened in the past and what could happen in the future in these situations?

Refer to the list of input and output devices on Student's Book pages 60–65 and 69–73. Which ones have not been mentioned in this lesson? How could they be used or incorporated into a system today?

Homework activity: encourage students either to ask family members about systems they have had access to or to research additional ones using the internet. They should then break them down into terms similar to those used in this lesson.

Assessment suggestions	The diagrams produced in **Activity 2** could be collected and assessed using the sample answers provided in **PPT2_4b** as comparison.

2.5 An introduction to presentations

Learning aims
- Combine text, image(s) and numeric data.
- Create presentation slides, including text, and images.

Differentiated learning outcomes
- **All students must** be able to create a basic slide and identify some basic presentation principles.
- **Most students should** be able to create a new presentation using text and graphics, experiment with design layouts and describe presentation terminology.
- **Some students could** create a new presentation using a full range of text and graphic tools, explain a full range of associated terminology and use design themes effectively to suit the styles of different organisations.

Resources
- **Student's Book:** pages 78–82
- **Files:**
  PPT2_5a
  WS2_5a

Starter suggestions

Discussion: Why do organisations use presentation software? Use Q&A or a mind map on the whiteboard to note down students' suggestions for possible uses. Answers should cover:

- presenting ideas to a large or small audience
- being able to show text, images and video to support a speech
- showing a diagram or set of instructions.

If possible, show an example of an *Apple* keynote presentation – *Apple* are famous for their use of visual presentations, for example, when launching products. Discuss how the speaker uses text, graphics and effects to complement what they are saying and not distract the audience.

Outline the lesson objectives: creating a basic slide and understanding key presentation terminology.

Main lesson activities

Student task: Ask students to carry out **Activity 1** on Student's Book page 78, using **PPT2_5a** to note down examples of presentation software and their features. When students have finished, discuss their answers and ensure that all the major software packages have been discussed.

Point out to students the information featured in the Real World box about the continuing development of Open-Source software that is free to use and distribute.

Discussion: Using your chosen presentation package, demonstrate and discuss each of the features outlined in the Student's Book on pages 79–80: slide, animation, transitions, slide master, design themes, slide layout, slide sorter and outline view. Assure students that they will have the opportunity to use all of these features in future sessions.

Demonstration: Opening a new presentation, adding text and shapes, changing fonts and shape properties.

Student task: Creating a basic slide. Ask students to carry out **Activities 3 to 5**, referring to the Student's Book as required. As the following sessions go into much more detail about producing presentations, the aim of this exercise is to simply experiment with some of the tools and features, especially text and shapes. Recreating the school logo (or any logo) is a good way to see how different fonts and shapes can give different impressions to the audience.

Demonstration: Using design themes. Show a variety of designs and ask students which of them would be suitable for a new exciting clothes shop, for example.

Student task: Set **Activity 6** and then ask students to show and describe which theme would be suitable for each of the types of organisation listed. If possible, present to the whole group. Students may well disagree (as design preferences can be very personal) and this could lead to useful discussion. Remind students that pre-designed themes can be useful for ideas and inspiration, but students should also create their own themes (see **Tip** on Student's Book page 82).

Give extra support by creating a screen recording of your demonstrations that students can refer to during the lesson. Free options include: *OBS Studio for Windows* and *the screen recorder included in QuickTime within macOS*.

Give extra challenge by asking students to create a more complex logo, maybe recreating a famous example or redesigning the school logo. Students confident with presentations could also present and demonstrate the section of the lesson concerned with terminology.

Plenary suggestions

Hold a Q&A session: Define some of the keywords from the session, including the various features of presentation software covered.

Discussion: Ask students to imagine they have to provide a presentation to a large audience about their school. What would be the five key pieces of information they would need to include? Responses might include: the logo and motto, a map of the building, pictures around the building, information on staff, students and course details.

Assessment suggestions	Homework: Ask students to watch some presentations online, for instance on the *Apple, Microsoft* and *TED* websites, and write a short report on how slides are used to complement what is being said. Ask students to investigate Open-Source software and the advantages and disadvantages of using it to both the general public and businesses.

2.6 **Presentation** preparation *(double lesson)*

Learning aims
- Combine text, image(s) and numeric data.
- Save and print documents and data.
- Use a presentation master slide to place objects and set styles.
- Create presentation slides, including text and images.

Differentiated learning outcomes
- **All students must** be able to create and save a new presentation and add text and colour to a master slide.
- **Most students should** be able to create and save a new presentation and create an appropriate master slide that uses text, graphics and colour.
- **Some students could** create, save appropriately and set the house style of a presentation, setting up a master page that combines appropriate text, styles, graphics and header/footer tools to produce a professional-looking template.

Resources
- **Student's Book:** pages 83–90
- **Files:**
  PPT2_4a.pptx
  PPT2_4b.pptx
  PPT2_5a.pptx
  PPT_6a
- **Further Resources:**
  Animation websites, such as:
  www.pixar.com
  www.aardman.com
  www.disney.com
  https://en.wikipedia.org/wiki/Twelve_basic_principles_of_animation

Starter suggestions

Discussion: How do students think animated films are made? Answers should include suggestions such as by the use of hand-drawn cartoons, models and miniatures, and modern computer-generated images. Basic animation could be shown using paper flick books; students could create them if time allows.

Refer again to the animated film studio as the basis of these practical sessions. Compare it with the following familiar companies: Pixar, Disney, Aardman, Industrial Light and Magic (ILM). Show their respective websites on screen and discuss the differences in animation styles:

- modelling and hand drawn work at Aardman
- traditional cell animation at Disney
- computer-generated imagery at ILM.

Outline the lesson objectives: setting up a presentation and creating a suitable house style.

Main lesson activities

Introduction: Refer back to the introduction to the task on page 58 of the Student's Book.

- Recap on the activity and structure of True to Life Studios, their core business of creating animated films but with a range of facilities and departments within the building.
- Refer to and show the system diagrams from Session 2.4 (**PPT2_4a** and **PPT2_4b**).
- Outline the practical task and the type of presentation to be made: a guide for new staff to the input, output devices and systems used at True to Life Studios.

Discussion: What is important in a professional presentation? Ask students to work in pairs or small groups and write down a list of ideas. These might include spelling, grammar, clarity, high-quality images, sensible colours, good quality logo, eye-catching, house style, use of animation and transitions. Make sure students understand the concept of a house style (see Student's Book page 83).

Make sure students understand the concept of house style and the importance of considering the audience when creating any presentation.

Student task: Ask students to carry out **Activity 1** on Student's Book page 83, make sure to discuss how language styles, types of imagery and the way information if presented should be adapted depending on the audience.

After a while, ask students to concentrate on the target audience of the presentation they will be creating, namely a business presentation for employees.

Demonstration: Use of master pages in *Microsoft PowerPoint*:

- setting backgrounds
- using fonts
- adding additional information such as a logo and header and footers.

Make sure you demonstrate how to move in and out of the master slide, even after information is added to slides. Use **PPT2_5a** as required for demonstrating skills.

Student task: Ask students to carry out **Activities 2 to 4**, referring to the information in the Student's Book as required.

Plenary suggestions

The first part of the double lesson ends with these activities. If a plenary is required, use peer feedback to comment on the presentations created so far.

Assessment suggestions	Homework: If the two parts of this double lesson are presented separately, students could be asked to create logo sketches for the animation company between lessons.

2.6 cont. Presentation preparation (double lesson)

Learning aims
- Combine text, image(s) and numeric data.
- Save and print documents and data.
- Use a presentation master slide to place objects and set styles.
- Create presentation slides, including text and images.

Differentiated learning outcomes
- **All students must** be able to create and save a new presentation and add text and colour to a master slide.
- **Most students should** be able to create and save a new presentation and create an appropriate master slide that uses text, graphics and colour.
- **Some students could** create, save appropriately and set the house style of a presentation, creating a master page that combines appropriate text, styles, graphics and header/footer tools to produce a professional-looking template.

Resources
- **Student's Book:** page 83–90
- **Files:**
  PPT2_6a.pptx
- **Further resources:**
  Animation websites, such as:
  www.pixar.com
  www.aardman.com
  www.disney.com
  http://en.wikipedia.org/wiki/12_basic_principles_of_animation

Starter suggestions

As this is the second part of a double lesson, a recap and introduction may be required.

Discussion: Refer back to the companies shown and discussed in the starter of the first lesson (Pixar, Disney, Aardman, Industrial Light and Magic (ILM)) and arrange students into groups to look at examples from each site.

- How have the different companies styled themselves in regard to colour and logo design?
- What sort of audience are they targeting?

Note down ideas and key terms on the whiteboard as they arise.

Main lesson activities

Student task: Ask students to sketch out a logo on paper for the True to Life Studios animation company.

Demonstration: How to insert *ClipArt* and use the drawing tools, text boxes and shapes within *MS PowerPoint*. It should be noted that some newer presentation packages have now removed traditional clipart and now include online graphic searches. Adapt the lesson accordingly.

Stress that it is important to choose images appropriate for the target audience. To reinforce this, show **PPT2_6a** (Slide 1) and ask students which graphics would be suitable for a business presentation. (Slide 2 shows which are appropriate and which are not.) What target audience would the other graphics be suitable for?

You should also encourage the use of drawing tools to create original graphics. Stress the importance of using different elements in combination rather than relying on one source.

Demonstration: How to group objects using the selection tool, and its use in graphical elements such as logos with multiple parts.

Student task: Ask students to carry out **Activities 5 and 6**, referring to the Student's Book as required. They should use *ClipArt, online graphics, WordArt* and *Autoshapes*, adding a logo to their master page and a heading to their first slide.

> **Give extra support** by asking those who create master slides reasonably easily to help those that have not. Some students could also be allowed to work without using master slides.
>
> **Give extra challenge** by asking students to create a more complex logo using image editing programs like those discussed in Session 2.5. Encourage students to develop additional skills, such as adding hyperlinks to navigate through the slides from the screen, and to access webpages relevant to animation.

Plenary suggestions

Ask all students to display their master slide on screen and all walk around the room giving peer feedback. Ask students to describe examples they have seen that look professional. Alternatively, worked examples could be uploaded to the school network and commented on in a more anonymous way.

Ask students to explain to the group why they think their layout is appropriate for the organisation.

Assessment suggestions	Homework: Students could spend time at home looking at the animation company websites shown in the first part of the lesson and write down specific devices and technology for each of the companies. Alternatively, they could describe their favourite and give the reasons for their preference.

2.7 Adding text, images and multimedia elements to a presentation
(double lesson)

Learning aims
- Combine text, images, animations, sound and video to create a presentation for a specific audience.
- Create notes for the presenter.

Differentiated learning outcomes
- **All students must** be able to add text and images to a presentation, most of which should be relevant.
- **Most students should** be able to add some relevant text and images to a presentation and use speaker notes to help the presenter.
- **Some students could** add images and descriptions for all the input and output devices in this unit and use speaker notes to explain the advantages and disadvantages of each.

Resources
- **Student's Book:** pages 91–96
- **Files:**
  PPT2_7a.pptx
  PP2_7b.pptx
  WS2_7a.docx
  Outline.txt
- **Further resources:**
  Software Clip Art

Starter suggestions

Discussion: Show the presentation in **PPT2_7a** and discuss the following:
- an appropriate amount of information – how much is too much or too little?
- font sizes – consideration of the audience and distance from display
- how to avoid distracting the audience by colour, images, etc.

Demonstration: Use the points raised during the discussion to build a better example.

Student task: Ask students to carry out **Activity 1**. If possible, give them access to **PPT2_7a** so they can also experiment, or ask them to create a similar example.

Outline the objectives of the lesson – see above.

Main lesson activities

Demonstration: Using a new blank presentation, show each of the different layouts available. Ask students to suggest uses for each and, where possible, add text or an image to demonstrate. Where full control of the layout is needed, it is best to use the **Title only** layout. Also, demonstrate how to create slides using a text outline. Use the file **Outline.txt** to create two slides with headings and subheadings.

Student task: Ask students to create a new blank presentation and experiment with layouts and inserting outlines. Either provide access to the **Outline.txt** file or ask students to create a similar text file. Their presentation should cover:
- input and output devices with descriptions, advantages and disadvantages.
- information on systems that use these devices (for example, the alarm, canteen, and shop).

Student task: Ask students to decide how many slides their work should need. Refer back to the brief as required. Students may find it useful to plan their presentation on paper, either using a storyboard (provided in **WS2_7a**) or by listing each slide and an idea of its contents (**Activity 2**, page 92).

The second part of the double lesson starts here. Most of this lesson should be given over to students working on their presentations. Keep the demonstrations short. Depending on time and how confident the group is, additional lessons could be inserted to give students time to add images and text to their presentation.	Demonstration: Using the images provided, insert and edit images, following the examples in the Student's Book if required (pages 93–94). The presentation created in class so far could be left running in a 'Loop' on screen. Student task: Tell students to work through **Activities 3** and **4**, adding images to their presentation. Demonstration: Adding text boxes and text, formatting the information and using speaker notes (Student's Book pages 94–95). Also show **PPT2_7b** to demonstrate how speaker notes should be used to remove unwanted text from a slide. Demonstrate how you can use **Presenter tools** to send slides to a projector. Student task: Ask students to complete **Activity 6** on page 95 of the Student's Book, adding descriptions to each device and speaker notes. Demonstration: Using the sound and video clips provided, insert a video onto a new slide and insert a sound as an individual file and then to accompany a transition. Student task: Ask students to carry out Activity 7 on Student's Book page 96. Students should investigate the different file formats available (see the Tips box on page 95) and source two files to put into their presentations. If time allows, recording quick video clips of devices in school would be ideal along with some basic editing techniques. Student task: Ask students to carry out Activity 8 on Student's Book page 96. They should investigate the different file formats available (see the Tips box on page 97), including sound files. If time allows, recording quick audio clips of devices in school would be ideal along with some basic editing techniques.

> **Give extra support: Soundbible** offers a wide selection of free sound effects and **Pixabay** offers free video clips.
>
> **Give extra challenge** by allowing students to experiment with audio editing using Open-source software such as **Audacity**) and free video editors such as **Windows Movie Maker**.

Plenary suggestions

Ask students to present one of their slides and ask for peer opinions. Are the images clear? Can the text be read? Is there too much? Or too little? What is the overall effect?

Alternatively, use the poor example in **PPT2_7b** and allow students to spot the obvious problems with the slides in the relationships between text and images.

Assessment suggestions	Students could create a storyboard of their presentation as homework. A new presentation on a hobby or favourite topic could be planned out using a storyboard and the guidance from this session.

2.8 Adding animation, transitions and additional elements

Learning aims
- Create a business presentation that combines text, images and graphics.
- In the presentation, use transitions and animation appropriate for a defined audience.

Differentiated learning outcomes
- **All students must** be able to use basic transitions and animation in their presentation.
- **Most students should** be able to use transitions and animation suitable for a defined audience and use drawing tools to describe ICT systems.
- **Some students could** create detailed diagrams to explain how ICT systems work and use animation and transitions for an adult business audience.

Resources
- **Student's Book:** page 97–102
- **Files:**
  PPT2_8a.pptx
  PPT2_8b.pptx
  PPT2_8c.pptx
  Sample_Data.xlsx
  WS2_8a.docx

Starter suggestions

Discussion: Show **PPT2_8a and PPT2_8b** (using **Slide show** mode) and discuss the differences between the two presentations. It should be clear that **PPT2_8a** uses suitable animation and transitions, whereas those used in **PPT2_8b** are unsuitable. (Make sure students understand why they are unsuitable.).

Outline the objectives of the lesson – see above.

Main lesson activities

Much of this lesson should be given over to students working on the presentations they started in Session 2.6. Many students may still be completing work from the previous lesson.

Discussion: Refer back to the input, process and output diagrams created in **Session 2.1** and **Session 2.3**. Remind students that the idea of their presentation is to display the devices used and describe the information processed.

Student task: If they haven't already sketched or created these diagrams on the computer, ask them to sketch diagrams to represent the three systems described in the Student's Book on page 97 (**Activity 1**).

Demonstration: Show how to use the drawing tools to create shapes, lines and arrows. The **Format Shape** option (usually available by right clicking a shape) can be used to set styles and colours. Many of these may have already been covered in **Session 2.5** and **Session 2.6**.

Student task: Ask students to carry out **Activity 2** on Student's Book page 97, creating system diagrams using drawing tools.

Demonstration: How to add a chart or graph and customise it. Create a chart using the **Insert > Chart** option and choosing a suitable chart, similar to that in the Student's Book. After using the **Insert** command, a sample spreadsheet will open. Use the *Microsoft Excel* file **Sample_Data.xlsx** and either copy or paste in some sample data. Areas can be edited by right or double-clicking different elements.

Student task: Ask students to carry out **Activity 3** on page 98 of the Student's Book. The sample data in this lesson should be made available to students. They should also be free to create their own data if they prefer.

Demonstration: Refer back to the examples in the starter and then demonstrate how to add animation to particular objects or text and then slide transitions. Use **PPT2_8c** if required (**PPT2_8a** with all animation/transitions removed). Remind students about appropriate use: there are lots of fun effects and styles that would be great for a children's party presentation but not a business adult audience. Refer students to the paragraph on 'Suitability and consistency' on page 100 of the Student's Book. Make sure you show students how to remove unwanted effects.

Student task: Ask students to carry out **Activities 4 to 6** on pages 99–100 of the Student's Book, experimenting with animation and transitions within their presentations (or a new blank file if desired) to enhance their work. After a short while, ask students to work in pairs to note down which choices would be most suitable for an adult audience, and then discuss answers as a class group.

Demonstration: Use any of the presentations so far to show how to change the order of slides using the **Slide Sorter**. Students may have designed their work in the correct order as they progress, but it is important they can use the feature. Also show the **Set Up Show** option – students will most likely use the default options, but explain the choices, especially **Presenter View** and its link to speaker notes. Then ask students to try **Activity 7** on page 101.

Any remaining time should be spent working on presentations.

Give extra support by providing chart data and sample system diagrams to those who require it.

Give extra challenge by asking students to animate their system diagrams.

Plenary suggestions

Peer assessment: Ask students to move around the room offering constructive criticism about other presentations. If space is limited, copies of the presentations could be uploaded onto the school network for peers to access (or USB memory sticks could be used).

Hand out **WS2_8a**, which contains a 'Presentation checklist', and ask students to make sure they have all the essential elements so far.

Outside visitor: Ask an adult or student not familiar with the project to come and offer evaluative comments.

Assessment suggestions	Students should compare their work so far to the checklist in **WS2_8a**, making sure they have included as many of the elements as possible. Elements cannot be ticked until the next lesson.

2.9 Finalising and displaying your presentation

Learning aims
- Save and print documents appropriately, creating notes for the presenter and audience.

Differentiated learning outcomes
- **All students must** be able to remove obvious errors from their work and print out their presentation.
- **Most students should** be able to check their presentation for text and graphical errors and print it in a suitable format for business use.
- **Some students will** ensure their work is free from all errors. They will print it in appropriate business formats.

Resources
- **Student's Book:** pages 103–105
- **Files:** WS2_9a.docx

Starter suggestions

Discussion: Using the whiteboard, create a mind map with the phrase 'Is my presentation ready?' at the centre. Ask students to contribute different checks that can be made and rules to follow. Suggestions should include things like:

- spelling
- grammar
- graphical errors
- accurate information
- clear images and graphics
- suitability for the target audience (adult business audience)
- checks made for animation clashes
- checks made for constant transition effect.

Students should know these by now but recap the important elements.

Main lesson activities

Demonstration: Show how to use the spelling and grammar tools within the presentation software. It is important to point out that the language setting should always be checked as the default setting may not be correct.

Demonstration: Graphical errors. Use one of the presentations to add layout errors, for example by overlapping shapes and text. These need to be checked by eye.

Student task: Ask students to carry out **Activity 1** (Student's Book page 103), checking for spelling, grammatical and graphic errors. If time allows, ask students to swap with a classmate so they can check each other's work and make any changes that are required.

Demonstration: Show how to save a presentation using compression settings by selecting the **Tools** option when using the **Save As** command. Discuss how lowering the ppi setting decreases the file size but also lowers the quality. When only displayed on screen, the effect will be minimal, but if printed, especially as full-page slides, the images will lose quality.

Student task: Ask students to complete **Activity 2** (Student's Book page 104) saving their final presentation. Remind students that if compressed, a presentation and its images cannot be returned to their original state. For that reason, it may be wise to save more than one version of the file.

Demonstration: Show how to print a presentation and discuss the various options for printing. Layouts are shown in the Student's Book, but it may be worth printing a selection and demonstrating using a multi-slide presentation. The number of slides per page will depend on the reader requirements, but if the speaker notes are required, then only the **Notes** layout will display them clearly.

Student task: Ask students to complete **Activity 3**. Presentations should be printed in **Notes** view if speaker notes have been provided.

> **Give extra support** by creating a prompt or cheat-sheet for the skills in this session.
>
> **Give extra challenge** by asking students to investigate the best compression settings to use in different scenarios.

Plenary suggestions

Bring the project to a close using the checklist in **WS2_9a** to recap the skills covered.

Ask for volunteers to present their work in a business style to the class.

Assessment suggestions	The quality of work should be judged as an adult business presentation with appropriate language. The checklist in **WS2_9a** could be used as both a student check-in sheet and an aid to teacher assessment.

3.1 Storage devices and media *(double lesson)*

Learning aims
- Identify a range of storage devices and media and their typical uses.
- Describe the comparative advantages and disadvantages of using different devices.
- Describe the difference between main/internal memory and backing storage.

Differentiated learning outcomes
- **All students must** be able to identify a range of storage devices and media.
- **Most students should** be able to identify a range of storage devices and media, their uses and the relevant advantages and disadvantages of each.
- **Some students could** explain the relevant advantages and disadvantages of a full range of storage devices and media depending on the needs of their user.

Resources
- **Student's Book:** pages 110–114
- **Files:**
  WS3_1a
- **Further resources:**
  Selection of storage devices (and one non-storage device).

  www.howstuffworks.com

Starter suggestions

Have a selection of storage devices out on display at the front of the classroom, for example:

- DVD/CD discs
- hard disk drive
- CD/DVD writer
- magnetic tape.

Include an object that doesn't belong to the category of storage devices, such as a computer mouse.

If one is available, an old personal computer is ideal for such demonstrations as the original source of devices is clearer to students. Additionally, any older storage devices – such as floppy disks, punched cards or zip/Jaz disks – will help to show product development.

Discussion: Ask students to identify the different storage devices and ask:

- What are the similarities between them?
- How are they used?
- Which device does not belong in this category?

Write comments on the whiteboard.

Give an outline of the lesson: storage devices and media – their uses, advantages and disadvantages.

Student task: Ask students to carry out **Activity 1** on page 110 of the Student's Book, writing down all the devices they have at home that can store information. Come back to this list after the first main lesson activity.

Main lesson activities

To prepare for the first class activity, use **WS3_1a** to create 12 cards, each showing the name of a different storage device. Divide the class into small groups and give each group one or more cards. You can either give out all the cards or use just a selection depending on time constraints.

Student task: Give students 20 minutes to produce a brief verbal presentation of their device (or devices) to present to the group. Their presentation should include: a description of the device, its advantages, disadvantages and possible uses. Students can gather information from the Student's Book and the internet (**www.howstuffworks.com** is a useful website).

Class activity: Watch as many presentations as time allows. After each one, pull together the group's comments, pointing out any errors or omissions from the Student's Book. Expand on their comments as necessary. Also prompt the class to ask questions about each presentation if appropriate.

Discussion: Refer back to the lists of devices at home written in **Activity 1**. How can these be developed and how has student understanding developed?

The first part of the double lesson ends with this class activity.

Plenary suggestions

Hold a Q&A session: Look at the devices listed in the Student's Book (pages 110–112). Were they all covered or discussed during the lesson activities? If not, which ones were not? These should be researched as homework.

Assessment suggestions	**Activity 1** could be extended as homework between the first and second parts of the lesson (if the two parts are separate), by students asking family and friends about the devices they own.

3.1 cont. Storage devices and media (double lesson)

Learning aims
- Identify a range of storage devices and media and their typical uses.
- Describe the comparative advantages and disadvantages of using different devices.
- Describe the difference between main/internal memory and backing storage.

Differentiated learning outcomes
- **All students must** be able to identify a range of storage devices and media.
- **Most students should** be able to identify a range of storage devices and media, their uses and the relevant advantages and disadvantages of each.
- **Some students could** explain the relevant advantages and disadvantages of a full range of storage devices and media depending on the needs of their user.

Resources
- **Student's Book:** pages 110–114
- **Files:**
  PPT3_1a.pptx
  WS3_1b.docx
  WS3_1c.docx

Starter suggestions

If the two parts of the double lesson are separate, a recap and introduction will be required. Refer students back to the assessment objectives.

If **Activity 1** was set as homework in the first part, you could start with a discussion about what students came up with (see also **Answers to Activity 1** in the previous lesson).

Main lesson activities

Discussion: Outline the following terms with respect to memory. They may come up in device descriptions:

- main or internal memory
- backing storage
- serial or sequential access
- direct/random access
- access speeds.

Make sure students are aware of the difference between memory (RAM/ROM) and storage.

Discussion: Using the storage capacity table outlined on page 113 of the Student's Book, discuss how different devices are designed for different needs (this will be covered later). Make it clear that the capacities given in the table are true at the time of printing – as technologies develop so do capacities.

Student task: Give students **WS3_1b** to help them carry out **Activity 2** – a small survey to investigate how many electronic files they own. Students may ask peers, or the task could be extended into a homework task, see below. Exact figures are not essential; the purpose is to help quantify files sizes. Where possible, ask students to add in file types, so that commonalities can be discussed.

Use the tables on page 113 in the Student's Book (or more recent figures if available) to calculate approximate storage requirements for each person surveyed. Students can then select a suitable storage method.

Discussion: Show **PPT3_1a** and discuss how the user's need should help decide on the most suitable device when storing information. Explain ease, speed, capacity and permanence.

Student task: Ask students to carry out **Activity 3** on page 114 of the Student's Book using **WS3_1c**. Students should work in small groups to consider the situations given and note down their ideas. Discuss each group's work.

> **Give extra support** by making sure you circulate through the groups, identifying and remedying problems as required.
>
> **Give extra challenge** by asking students to investigate the future, and implications, of storage devices and media, the increasing use of cloud storage and cutting-edge systems such as quantum storage or holographic layers.

Plenary suggestions

Discuss the **Real World** boxes on pages 112 and 113 of the Student's Book.

Hold up the selection of devices from the beginning of the first lesson and ask students to suggest with some further real-world uses. Question students as to why other devices would not be as suitable. Also, ask students to consider how these devices could develop in the future.

Recap the objectives of the lesson.

Assessment suggestions	**Activity 2** could be extended to home and friends, with students carrying out a larger survey of files and file types. This could be developed by asking students to calculate averages.

3.2 The importance of data backup

Learning aims
- Define the term 'backup' and describe the need for taking backups.

Differentiated learning outcomes
- **All students must** be able to give a basic description of the term 'backup' and suggest why it is important.
- **Most students should** be able to describe the term 'backup', state its importance for different organisations and outline some of the different methods available.
- **Some students could** explain the term 'backup', describe the different models and solutions available and make suggestions based on the needs of their user.

Resources
- **Student's Book:** page 115–116
- **Files:**
  PPT3_2a.pptx
  WS3_2a.docx
  WS3_2b.docx
- **Further resources:**
  Backup software list at https://en.wikipedia.org/wiki/List_of_backup_software

Starter suggestions

Student task: Ask students to carry out **Activity 1** on page 115 in small groups, listing the information normally stored at home. Ask them to consider what would happen if the information were lost and what could be done to replace it.

Discussion: Ask students how many of them actually have personal information backed up and if so, how. Use any real-life examples you know of data being lost - students may have their own examples. How is student data backed up at your school? What happens when they leave? Make sure students are confident before moving on that they understand the concept of a backup copy. Include terminology like 'backup frequency', 'cycles' and 'archiving'.

Give an outline of the objectives of the lesson: the importance of backing up information in any situation and the different methods available.

Main lesson activities

Student task: Ask students to carry out **Activity 2**, filling in the first two columns of the table in **WS3_2a**. This could be done individually or in groups.

Discussion: Create a mind map on the whiteboard of the three situations in **Activity 2** and around them add student suggestions. Ensure that students are aware of the level of importance attached to different information. A school essay lost may be inconvenient, but the loss of a hospital patient's record history could result in incorrect treatment.

Discussion: Use **PPT3_2a** (Slide 1) as an on-screen prompt; discuss the four models of backup shown. Although not specifically in the specification, knowledge of them will certainly aid understanding of the process. This slide could be printed and handed out for students to make notes on during the discussion.

Discussion: Use **PPT3_2a** (Slide 2) to discuss different backup solutions. Again, students could make notes on a printout of it. Rather than repeating the last activity, students could work in groups to research and present information on the four solutions, using the Student's Book and internet as a guide.

Discussion: Describe how operating systems have inbuilt backup systems, such as Windows Backup and Apple Time Machine (Linux has multiple options). There are also third-party solutions – the term 'third party' will need clarifying. Explain that the benefit of a software solution is that backing up of data is done automatically and on a regular basis, so the user can work with confidence.

Student task: In groups, students complete **Activity 3** (Student's Book page 116) using **WS3_2b**. Students will need access to the internet as the results will depend on current software (see the Wikipedia link in **Further resources**). Specific software information is not important, just the concept of automated backup tools.

> Give extra support by making sure you circulate among students, identifying and remedying problems as required.
>
> Give extra challenge by asking students to return to **Activity 2** to add suitable devices, backup models and solutions to the four scenarios in **WS3_2a**.

Plenary suggestions

Discussion: Ask students if their opinion on backing up information has changed during the session. What will they do differently now? What backup regime is most appropriate for a student today?

Discussion: Make sure students read through the **Real world** and **Language** boxes on pages 115–116 of the Student's Book.

Mini quiz: Select key words from the Student's Book and glossary to test students' knowledge.

Discussion ideas: What are the advantages and disadvantages of the following:

- online storage and cloud computing
- police storing criminal records
- owning personal music and video only in electronic form?

Assessment suggestions	Answers to **Activity 2** could be taken in for assessment. If so, ask students to give as much detail in their answers as possible using the terms outlined in this session.
	A homework activity might be to speak to a family member or friend and ask how the organisation they work for backs up its important information. This could be presented as an example to the class in a future lesson.
	Ask students look again at the organisations in this session and consider the security implications of the information they store. Which of them would benefit from encrypting their data?

3.3 Banking facilities today

Learning aims

- Demonstrate an understanding of ICT in everyday life in terms of batch, online and real-time processing.
- Describe the use of banking applications including EFT, ATM, bill payments, credit/debit cards, cheques, telephone and internet banking.
- Demonstrate an understanding of the wider-world applications for electronic, internet and mobile communications.

Differentiated learning outcomes

- **All students must** be able to identify a range of modern banking facilities and describe some of the technology used.
- **Most students should** be able to describe a range of banking facilities and methods, as well as some of the technologies and communication systems behind them.
- **Some students could** explain how modern banking facilities, methods and technologies are used with an understanding of the communication applications linked to them.

Resources

- **Student's Book:** pages 117–118
- **Files:**
  WS3_3a.docx
  WS3_3b.docx
  PPT3_3a.pptx
- **Further resources:** Projector, sticky notes.

Starter suggestions

Student task: On the whiteboard write the term 'Banking technology' and hand out sticky notes to all students. Ask them to write on the notes an example of how they imagine computers are used in banking today and stick them on the board. These can then be referred to again at the end of the lesson. If students initially struggle, show **PPT3_3a** (Slide 1) for a selection of current uses they might link computers to.

Discussion: Show **PPT3_3a** (Slide 2) and outline the main technologies that will be covered in the lesson. Also remind students that many of the input and output devices used are also detailed in Unit 2. These systems will also use storage devices outlined in Sessions 1 and 2.

Main lesson activities

The majority of the time in this lesson should be given to students to work on **WS3_3a**, which contains a table covering the key aspects of how ICT is used in modern banking. **WS3_3a** is not listed as one of the activities in the Student's Book, but it allows students to produce guided notes.

Discussion: Provide a basic description of each of the facilities and technologies shown in **PPT3_3a**. Ensure that all students understand the basic concept of each. Encourage students to make notes if required.

Student task: Either individually or in small groups, students should complete **WS3_3a**. They should use the Student's Book and the internet as required. At regular points, stop the class and ask for contributions, such as examples of why cheques are still used and what the advantage of an ATM machine is. This will help identify where help is needed. The completed worksheets should be handed in for assessment.

Give extra support by explaining difficult keywords and concepts individually or by making sure students are aware of the assistance in the Student's Book.

Give extra challenge by asking students to refer back to Unit 2 and include input, output and storage devices in their explanation of banking methods, data storage and backup.

Plenary suggestions

Discussion: Look back at the sticky notes on banking technology created in the starter, removing duplicates and identifying those covered in the lesson. What technologies were not discussed? Are they relevant?

Discussion or student task: Complete **Activity 1** on page 118 using **WS3_3b** – linking methods and technology to different scenarios.

Discussion: **Activity 2** (Student's Book page 118) – What are the security risks linked to electronic banking methods? Search for online banking fraud on the BBC News site and display to the class.

Discuss the different methods of banking fraud today, such as phishing emails and websites, trojan horse viruses (using key stroke logging) and false ATM fronts.

Student task: **Activity 3** on page 118 – Ask students to list the types of information that would need to be stored in the banking facilities described in this session.

Assessment suggestions	Collect in **WS3_3a** and assess. Answers should be based on collecting information from the Student's Book and additional information from the internet.
	If time is short, **Activity 3** could be carried out as homework, as it links into the following practical sessions.
	How are banks using methods like two-step verification to make access to their online facilities more secure?

3.4 Document preparation

Learning aims
- Combine text, image(s) and numeric data.
- Save and print documents and data.
- Set page layout options including size, orientation, margins and columns.
- Use the header and footer of a document to add appropriate information.

Differentiated learning outcomes
- **All students must** be able to create and save a new document and add some information.
- **Most students should** be able to create and save a new document, edit the page settings and add a header and footer.
- **Some students could** create, save appropriately and set the house style of a document, including page, margin and columns settings, also adding appropriate header and footer information to suit the style of the document.

Resources
- **Student's Book:** page 121–125
- **Files:**
  PPT3_4a.pptx
- **Further resources:**
  Banking websites:
  www.barclays.com
  www.hsbc.com
  www.db.com

Starter suggestions

Discussion: Refer to the banking company Abacus International introduced on page 108 of the Student's Book (and used as the basis of these practical sessions). Find a selection of local or international banks and project their websites to the class (examples are given in **Further resources** above). Alternatively, collect promotional printed documents from a range of local banks to show or build a display around. What sort of image are the banks putting across through their graphic styles and presentation? What is important to those creating them?

Student task: Ask students to work in pairs or small groups and write down a list of ideas. These might include:
- excellent spelling and grammar
- clarity and professionalism
- use of high-quality images
- suggestion of money and wealth
- sensible colours
- good-quality logo
- eye-catching design.

Outline the session objectives: setting up a document and creating a suitable house style. Check that students are familiar with the term 'house style.'

Main lesson activities

Discussion: Focus on the practical task and the type of document to be made, that is, a training guide for staff involved in backing up and storing data. The example shown in the Student's Book uses *Microsoft Word (Microsoft 365),* but a similar program could also be used.

Demonstration: Show how to create folders, sub-folders within them, and how to edit their names. Remind students about the importance of sensible and coherent file and folder names that clearly identify contents, especially in the professional world.

Student task: Ask students to carry out **Activity 1** on page 122 of the Student's Book.

Demonstration: Show how to set the following document properties: paper size, orientation, margins and headers and footers. Make sure you demonstrate the full range of options for the header and footer. The Student's Book shows text, date information, page numbering and file location. These should suit your local requirements and printer availability. The student's document could either be portrait or landscape, but portrait will probably be more suitable.

Student task: Students should carry out **Activities 2** to **5**, referring to the Student's Book as required.

Discussion: Refer back to the companies shown and discussed in the starter. How have they styled themselves in terms of colour and font choices? Students will have more opportunity to style their own document in later lessons.

Demonstration: Show how to add columns to a document. Use **PPT3_4a** to give an example of a possible layout. If possible, collect a few newspapers or magazines before the lesson to show real examples of how document pages are laid out. Demonstrate how to add columns – most students will probably use two. Point out that columns cannot really be seen until text is added – it might be beneficial to add sample text and ask students to do the same.

Student task: Students should carry out **Activity 6** on page 125 of the Student's Book.

Give extra support by asking students who create their template quickly to help others. Provide an annotated example of a printed template that students could quickly follow, showing label sizes, footer details, etc.

Give extra challenge by asking students to create a complex graphical logo for the bank rather than a text-based one. Other shapes and graphics could also be added to the header and footer.

Plenary suggestions

Ask students to explain to the group why their layout is appropriate for the organisation.

Show **PPT3_4a** again on screen and ask students to check they have included all the relevant information.

| Assessment suggestions | Students could be asked to design a logo for the organisation before the next lesson. |

3.5 Adding text and images to a document
(double lesson)

Learning aims
- Create a document that combines appropriate text and images and appeals to an adult professional audience.

Differentiated learning outcomes
- **All students must** be able to add text and images to a document, most of which should be relevant.
- **Most students should** be able to combine relevant text and images to a document that includes descriptions of storage devices and media.
- **Some students could** combine relevant images and descriptions of all the storage devices and media in this unit to create a professional-looking document, clearly explaining the advantages and disadvantages of each.

Resources
- **Student's Book:** pages 126–130
- **Files:**
  PPT3_5a.pptx
  Abacus Leaflet.docx
- **Further Resources**
  https://issuu.com/
  (A free online archive of printed publications)

Starter suggestions

Discussion: Before the lesson, collect a range of leaflets and flyers from various sources (including banking ones if possible) and lay them out as a display (If examples cannot be found, the website issuu.com listed above can be searched for banking material).

Student task: Ask students to write a list of key points they should consider when designing for an adult audience – specifically the bank described in this unit. Answers may include: formal style, clear, stylish, simple but not childish, uncluttered.

Outline of the lesson: working with text and images and incorporating them into a leaflet.

Main lesson activities

Demonstration: Using a new blank document, demonstrate how to format text using fonts, style, alignment and size. Follow the examples in the Student's Book if required. Use **PPT3_5a** to illustrate the difference between serif and sans-serif fonts. Refer to the examples from the starter task to see how they are used.

Student task: Students should carry out **Activity 1** on page 127 of the Student's Book, experimenting with modern and traditional styles of text. Allow students time to discuss and make choices for their own document.

Student task: Ask students to look at all the notes they made in Sessions 3.1 to 3.3 and the information in the Student's Book to help decide how many pages their training document may need. Refer them back to the brief on page 112 as required.

Demonstration: Show how to use bullets, paragraph formatting and tables. Use sample text or the file **Abacus Leaflet.docx** as required. Make sure students are happy with the following concepts:

- line spacing
- rows and columns
- cell alignment
- indentation.

Either provide a piece of sample text or ask students to use text they already have.

Student task: Encourage students to experiment with the options for text and tables and then carry out **Activity 2** (Student's Book page 129), writing an introduction for the training document created in the previous lesson. Again, make sure students refer back to the project brief in this unit if they need to (Student's Book page 112).

Plenary suggestions

Print and present a sample of students' work and have it peer-assessed, asking for comments on layout, use of formatting and clarity. Ask students to write on suggestions for improvement and, if possible, allow time for students to improve their versions.

Assessment suggestions	Homework: **Worksheets 3.8** and **3.9** can be used to collect information on storage media and devices between lessons. **Activity 2**, writing an introduction, could be given as homework.

3.5 cont. Adding text and images to a document (double lesson)

Learning aims
- Create a document that combines appropriate text and images and appeals to an adult professional audience.

Differentiated learning outcomes
- **All students must** be able to add text and images to a document, most of which should be relevant.
- **Most students should** be able to combine relevant text and images to a document that includes descriptions of storage devices and media.
- **Some students could** combine relevant images and descriptions for all the storage devices and media in this unit to a professional-looking document, clearly explaining the advantages and disadvantages of each.

Resources
- **Student's Book:** pages 126–130
- **Files:**
  PPT3_5a.pptx
  Abacus Leaflet.docx
  WS3_5a.docx
  WS3_5b.docx
- **Further Resources**
  https://issuu.com/
  (A free online archive of printed publications)

Starter suggestion

If this session is being delivered as two single lessons, a recap and introduction will be needed at the start of the second lesson.

Main lesson activities

The majority of the second part of the lesson should be given over to students working on their documents, keeping the essential demonstrations short. Depending on the length of the lessons and the group, additional lessons may be needed for students to add all the necessary information to their documents.

Students should already be familiar with internet searching and **WS3_5a.docx** and **WS3_5b.docx** are available to record information if required. Students will probably need extra time to complete these. They could be given as homework between the lessons or as an extra lesson, depending on the lesson length and the group.

Demonstration: Inserting an image – this is covered in detail in Unit 2 and the process is the same for all office programs. If required, refer to Session 6 of Unit 2.

Student task: Ask students to carry out **Activity 3** (Student's Book page 129) adding images to their work. Students should be reminded about not overcrowding their work and leaving space for text. As a group, discuss the most suitable number of devices to place on each page.

Demonstration: Use the file **Abacus Leaflet.docx** to demonstrate how text and images can be combined and the various choices available. Make sure to demonstrate all features from **Wrapping** to **In front of text**, as each feature has advantages and disadvantages. The **In front of text** option, for example, allows images to be moved much more freely, but space has to be made. Also demonstrate how to add hyperlinks and bookmarks, explaining the term hyperlink and how it can appear in all types of document as well as web design.

Student task: Students should now carry out **Activity 4** (Student's Book page 130), adding descriptions, advantages and disadvantages to each device. Information should be used from previous lessons and the Student's Book if required.

Give extra support by providing a selection of images in order to save time.

Give extra challenge by allowing students to use image-editing software to cut out or edit their images.

Plenary suggestions

Print and present a sample of students' work and have it peer assessed, asking for comments on layout, use of formatting and clarity. Ask students to write on suggestions for improvement and, if possible, allow time for students to improve their versions.

Assessment suggestions	Homework: **WS3_5a and WS3_5b** can be used to collect information on storage media and devices between lessons.

3.6 Creating a simple data model and chart

Learning aims
- Enter and edit data from different sources.
- Organise the page layout including layout, size and the use of headers and footers.
- Show an understanding of the use of basic spreadsheet modelling in real-life situations.
- Create a spreadsheet model layout.
- Produce a chart and insert it into another document.

Differentiated learning outcomes
- **All students must** be able to create a new spreadsheet file and add a title. They should be able to list useful spreadsheet functions.
- **Most students should** be able to create and save a new spreadsheet, add a title and set some page layout options. They should also describe formulae and functions that could be used in their work.
- **Some students could** create and save a spreadsheet model with an appropriate name and file location and add suitable header and footer details. They will also be able to explain appropriate formulae and functions to meet the needs of the brief.

Resources
- **Student's Book:** pages 131–137
- **Files:**
  SS3_6a.xlsx
  Abacus Leaflet Chart.docx

Starter suggestions

Students may be fairly experienced at creating spreadsheet models by now, so use your professional judgement to decide how much of this basic introduction is required depending on the group.

Discussion: The training document created in previous lessons should now be almost complete. In addition to text and images, students now need to add a chart that compares and displays the storage capacity of different devices.

Outline the objectives: An introduction to spreadsheet skills and creating charts.

Demonstration: Remind students about the importance of using folders and sensible filenames. Show how to open the spreadsheet software and create a new blank workbook. Then ask students to create a folder in their document area called SPREADSHEET MODEL for the work they do in this lesson.

Main lesson activities

Demonstration: Ideally with the spreadsheet software projected to the class, discuss the basic principles of modelling and how it can be used to test scenarios in theory before putting them into practice. Be sure to cover: software tool layout, rows, columns, cells and cell formatting, borders and the use of the formula bar.

Student task: Ask students to carry out **Activity 1** (Student's Book page 132). This could easily be done from the Student's Book and students asked for answers.

Demonstration: Show how to enter data into a model using a blank workbook. Demonstrate basic addition, subtraction, multiplication and dividing formulae using basic data. Also include changing column width and row height at this point.

Student task: Ask students to carry out **Activities 2** and **3 (Student's Book pages 132–133)**: a basic multiplication table. Remind them to save all their work into the folder set up at the beginning.

Demonstration: Setting page and layout, adding headers and footers, inserting and deleting rows and columns and merging cells.

Student task: Ask students to carry out **Activity 4** (Student's Book page 134).

Demonstration: Again projecting the software to the class, recreate the basic formula shown on page 134 of the Student's Book. Show how this can be replaced with a function. Ask students to study the diagram on page 134 (the structure of basic functions) and then demonstrate using the SUM function to add together the cells.

Student task: Ask students to carry out **Activity 5**, recreating the table on page 135. The Student's Book provides guidance on the content, but the activity could also be done as a group activity. Point out that the SUM function and quantities are not required to create the chart, but the model can also be used to calculate the total storage capacity of the number of devices in the table. If time allows, show students how to use borders and text formatting (although not a requirement of this session).

Demonstration: Show how to create a bar or column chart like the one on page 135 of the Student's Book. Use the spreadsheet file **SS3_6a** to demonstrate how to create a simple chart. Then show how each element can be edited and selected. Use example **Abacus Leaflet Chart** as required.

Student task: Ask students to carry out **Activity 6**: create a chart using the data from **Activity 5**. Allow them time to experiment with formatting and colour options. When selecting colours and styles, encourage them to consider the style of their training document from the last session.

Demonstration: Show how to **Copy** and **Paste** a chart into a written document. Demonstrate how to use the **Paste Special** option, allowing greater flexibility.

Student task: Ask students to carry out **Activity 7** (Student's Book page 137), adding the own chart to their training document.

> **Give extra support** by providing the data file for students to use.
>
> **Give extra challenge** by asking students to research accurate storage capacities of current devices. Experiment with other charts that provide the same visual information. Print and proofread documents in preparation for the next session.

Plenary suggestions

Discussion: What other charts could be used to present the same information?

Demonstration: Experiment with the table from **Activity 5**. What happens when the quantities are adjusted?

Assessment suggestions	The next session relies on the fact that the training document is complete. If possible, allow it to be completed as homework or provide out-of-hours computer access in school during break times or after school.

3.7 Final document checks and presentation *(double lesson)*

Learning aims
- Save and print documents, data and objects.
- Use proofreading and software tools to ensure that the documents are error free and corrected.
- Enter and amend text, numerical data, functions and formulae with 100 per cent accuracy.
- Demonstrate that the model works with appropriate test data.
- Adjust the display features in a spreadsheet.

Differentiated learning outcomes
- **All students must** be able to check their work for errors, test some of the elements of their model and print out all of their work.
- **Most students should** be able to use tools to check for errors, create tests to check their model and print out their work in suitable formats, including showing formulae.
- **Some students could** ensure their work is free from all errors and print it in suitable formats, including formulae. Tests should be used to check their model is accurate and their documents meet the needs of the brief.

Resources
- **Student's Book:** pages 138–143
- **Files:**
SS3_7a.xlsx

Starter suggestions

Discussion: Presenting professional documents means making sure there are no errors and testing data models thoroughly as commercial mistakes can be costly.

Give an outline of the lesson: document checks, proofreading and visual checks, spreadsheet testing, printing and file compression.

Main lesson activities

Demonstration: Show how to use the spelling and grammar tools, always checking the language setting, within their word processing and spreadsheet software.

Student task: Ask students to carry out **Activity 1** on page 139 of the Student's Book, checking for spelling and grammatical errors in both files.

Discussion: Why is it important to proofread documents? Printout their document from Session 3.6 and peer assess to look for any text or graphical problems. Refer to the list of potential checks on page 139 of the Student's Book.

Student task: Ask students to carry out **Activity 2** (Student's Book page 139), proofreading and acting on text and graphic changes that need to be made.

The first part of the double lesson ends after the student task based on **Activity 2**. If this session is being delivered as two single lessons, a recap and introduction will be needed at the start of the second lesson.	Demonstration: Spreadsheet testing and how to use simple data, rather than realistic data, to ensure formulae work correctly. Please note: The spreadsheet created in this unit is very simple, but the process of testing is still important. Use **SS3_7a**, also shown on page 140 of the Student's Book, to demonstrate. Larger spreadsheet models will require much more detailed testing. Student task: Ask students to carry out **Activity 3**, using their completed test table and spreadsheet to carry out simple tests make any changes that are necessary.

Demonstration: Show how to print the training document and spreadsheet. Discuss the various options for printing each type of document, including showing and printing formulae. Make sure to cover the skills needed to complete **Activity 4**.

Student task: Ask students to carry out **Activity 4** (Student's Book page 142), printing the training guide, the spreadsheet displaying values and the spreadsheet again displaying formulae with gridlines, row, and column headings all fitted to one page.

Please note: **Activity 5** is an extra desktop publishing activity to be completed if there is time, or as homework or a controlled assessment as described below.

Demonstration: Show students how to create PDF versions of their documents and explain how they can be viewed using a PDF viewer or internet browser.

Demonstration: Show students how to create a single compressed Zip folder from a selection of files. It is important that only the files required are placed into a suitable folder that has been correctly named so that the Zip file will also use that folder name.

Student task: Student should carry out **Activity 6**, first creating PDF versions of the files before creating a Zip folder of all the files they have created.

> Give extra support by allowing students to work in pairs or small groups when testing and printing.
>
> Give extra challenge by allowing students to complete Activity 5. Focus on the change of target audience (students); how might the document styling change?
>
> Send the final compressed Zip file as an email attachment to a peer in class.

Plenary suggestions

Bring the project to a close. Recap the skills and theory covered.

Ask for volunteers to present their work in a business style to the class.

Assessment suggestions	The additional banking information requested in Activity 5 may be more suited to homework or carried out in test conditions as a controlled assessment task. The printouts can be collected in for assessment. The training guide should be checked for technological accuracy as well as for its professional design.

4.1 An introduction to computer network hardware

Learning aims
- Describe a range of network devices and their purpose.
- Describe the use of WI-FI and Bluetooth in networks.
- Make decisions on the most appropriate network devices based on user's needs.

Differentiated learning outcomes
- **All students must** be able to identify basic network hardware and possible uses for them.
- **Most students should** be able to describe common network hardware and appropriate uses for them.
- **Some students could** explain the different uses for network hardware and the most appropriate hardware for a range of given situations.

Resources
- **Student's Book: page 148–151**
- **Files:**
  PPT4_1.pptx
  WS4_1a.docx

Starter suggestions

Collect and display a selection of network devices, such as cables, NIC cards, switches, etc. Ask students if they can identify them and suggest their purpose.

Give an outline of the lesson: network devices and their uses.

Discussion: Divide the class into four groups and assign each group one of the pieces of hardware described in the table on page 148 of the Student's Book, that is, NIC, hubs and switches, bridges, routers or modems.

Student task: Using the notes in the Student's Book and the internet, if available, ask each group to create a short presentation for their device. This could be electronic but a short verbal presentation would be fine. After five or ten minutes, ask groups to present what they have found. This should be a short activity, not a whole lesson.

Main lesson activities

Discussion: Compare the students' presentations to the information in the Student's Book and fill in any gaps or clear up any confusion.

Student task: Ask students to list the devices we use in the home that use either Bluetooth or Wi-Fi and why is one chosen over the other. Bluetooth speaker systems, for example, create a direct link with an audio source at a short range, whereas a wireless tablet may be needed to move around a home, and often garden.

Student task: Ask students to carry out **Activity 1** on page 149 of the Student's Book, writing down a list of reasons why a supermarket would find a router more useful than a bridge. Once students have finished, discuss answers with the group.

Discussion: Choosing between cables or wireless systems. Show **PPT4_1a**, slide 1, which summarises the factors to consider when making this choice. Talk through each one providing examples and then show slide 2 that extends the considerations to the type of device used to connect to a network.

Student task: Ask students to carry out **Activity 2** (Student's Book page 150). Students can use **WS4_1a** to note down their answers. As an extra task, students can add scenarios of their own choosing. Discuss their answers when complete and include as many devices as possible.

Discussion: Twisted pair cables in comparison to fibre optic cables. Outline each type of cable and their relevant advantages and disadvantages.

Student task: Ask students to carry out **Activity 3** (Student's Book page 150). Expand the discussion to include the use of cloud computing, especially the advantages and disadvantages of keeping files, documents and programs entirely online. Dispel the common misunderstanding about the use of the word 'cloud', this doesn't mean information is purely wireless or in the air, it just means an Internet connection is placed between the user and data - removing any notion of distance.

> Give extra support by using group work for the Student's Book activities.
>
> Give extra challenge by asking students to add some more scenarios to **Activity 2**.

Plenary suggestions

Return to the devices shown at the start of the lesson and discuss the question: how much more do students know now?

Quick Q&A session: Call out terminology from the session and ask students to define each item you mention.

Discussion: Where do students think these devices are being used? At home, school or in workplaces?

Assessment suggestions	Activity 3 could be given as homework and collected in for assessment. Ask students to design the layout of a network and try to include as many real devices as possible. This may include: <ul><li>recreating the network used in their ICT room at school</li><li>a home network</li><li>a network for use in a business environment, such as a computer store showroom.</li><li>incorporating cloud computing systems where appropriate.</li></ul>

4.2 Computer networks

Learning aims
- Describe a home network and the need for an ISP.
- Describe the need for browser and email software to take advantage of an internet connection.
- Define the terms Local Area Network (LAN), Wireless Local Area Network (WLAN) and Wide Area Network (WAN).
- Describe the differences between LANs, WLANs and WANs, identifying their main characteristics.
- Describe the characteristics and purpose of common network environments, such as intranets and the internet.
- Describe other common network devices (including hubs, bridges, switches and proxy servers).

Differentiated learning outcomes
- **All students must** be able to identify and discuss basic network systems.
- **Most students should** be able to describe LAN, WLAN and WAN systems and identify uses for them.
- **Some students could** explain appropriate uses for LAN, WLAN and WAN systems and also demonstrate an understanding of the technology involved.

Resources
- **Student's Book:** pages 152–155
- **Files:**
  PPT4_2a.pptx
  WS4_2a.docx
  WS4_2a_Answers.docx

Starter suggestions

Discussion: Ask students if they have heard of the terms LAN, WLAN and WAN and, if so, where?

Outside speaker: If possible, ask a member of the technical support team to come in and talk to the students about the systems in school. This could be as a starter or part way through the lesson.

Give an outline of the lesson: examining different network types.

Main lesson activities

Discussion: Show **PPT4_2a** (Slide 1) and outline LANs. Outline client-server (file, database and print servers) and peer-to-peer networks. Students may have heard of peer-to-peer sharing as it is often referred to in file-sharing applications. Such applications allow people across the world to share files on their computer and is therefore linked to software and media piracy.

Also briefly describe FTP (File Transfer Protocol) and its use in transferring files from one host to another – for example, uploading files on your computer to the internet.

Student task: Ask students to carry out **Activity 1** (Student's Book page 153) and complete the true/false questions using **WS4_2a**. (The 'Explain why' column can be given as an extra challenge – see below.) Discuss the answers with the group.

Next ask students to read through the advantages and disadvantages of LANs on pages 153 and complete **Activity 2**, again using **WS4_2a**. Discuss the answers using **WS4_2a_Answers**. Issues such as data corruption may need to be explained: data corruption can occur either by user mistakes (including multiple-user access to a file), through software problems or maliciously via a virus. Network security may also arise – this is covered in Session 4.7.

Discussion: How does the school network relate to the ideas discussed so far? Then show **PPT4_2a** (Slide 2) describing WLANs and discuss this type of network, using real examples. Point out that a WLAN is the same as a LAN but without the cables.

Activity 3 on page 154 could be carried out individually by students after they have read the information so far, or it could be done as a whole class activity, in which case use the whiteboard to record and discuss the answers.

Discussion: Show and discuss Slide 3 of **PPT4_2a** describing WANs. Outline how large organisations use WANs to link multiple networks across the world. This allows employees to work almost anywhere. Networks can be connected by cables, fibre optic networks and satellites.

Student task: Ask students to read through the advantages and disadvantages of WANs (Student's Book page 155) and complete **Activity 4**. Show Slide 4 of **PPT4_2a**. This illustrates the scenario of a bank with offices across the world. Give students time to work out how each LAN would be connected together in a WAN. When complete, show Slide 5 of **PPT4_2a**, which offers a possible answer.

Give extra support by providing a partly completed answer to Activity 4. Print out slides from **PPT4_2a** before the lesson to give as handouts. These could be annotated for clarity.

Give extra challenge by asking students to add explanations to the questions on **WS4_2a** or ask students to investigate how networks are employed in the following real-world examples:

- network security
- data farms
- online shopping
- the aerospace or automotive industries.

Plenary suggestions

If possible, show a variety of answers to Activity 4 on the whiteboard and discuss where students got it right and wrong.

Assessment suggestions	**Activity 4** could be given as homework and collected in. This could be extended by asking students to describe the issues (such as language and time zones) that may arise in creating a worldwide computer network.

4.3 The internet and intranets

Learning outcomes
- Outline the characteristics and purpose of common network environments such as intranets and the internet.
- Describe the use and purpose of common network devices (including hubs, bridges, switches and proxy servers).
- Understand the problems of confidentiality and security of data, including problems surrounding common network environments.

Differentiated learning outcomes
- **All students must** be able to identify and discuss modern communication methods and technology.
- **Most students should** be able to describe modern communication methods and technology including email, IM, video conferencing, blogs, social networks, facsimile and telephone applications, and describe business uses of each.
- **Some students should** be able to explain modern business uses for communication methods and technology including email, IM, video conferencing, blogs, social networks, facsimile and telephone applications.

Resources
- **Student's Book:** pages 156–158
- **Files:**
  PPT4_3a.pptx
  WS4_3a.docx
  WS4_3a_Answers.docx

Starter suggestions

Discussion: Briefly explain how an intranet works (this will be explained more fully later in the lesson). Ask students to list and discuss the possible features in both intranets and the internet: text, images, sound, video, and animation.

Give an outline of the lesson: how we use the internet and intranets, what the difference between them is and why both exist.

Discussion: How many students type in full website addresses and understand them? How many simply type what they want into Google (or another search engine) and select the first result? More information on internet searching can be found in Session 1.8 (Student's Book pages 37–42).

Discuss the use of sponsored links in search engines. Stress how important it is to check a range of results when searching, as commercial sites often try to make their way to the top of any search.

Write the address shown on page 156 of the Student's Book on the whiteboard:
https://www.harpercollins.co.uk/corporate/about-us/

Annotate the URL with students, explaining what each part represents. Also make sure to point out that not all website addresses will display a filename in the last column, this is due to the way modern websites are structured and the use of dynamic or hidden pages.

Student task: Hand out **WS4_3a** and ask students to complete **Activity 1** on page 156. Students should also find some additional examples of real websites and break down their URLs using the extra rows provided in the table on **WS4_3a**.

Main lesson activities

Discussion: Internet Service Providers (ISPs) and the three main ways of connecting to the internet. Connection is also possible via satellite, but this is rarely a practical solution. Divide the class into three groups and ask each to create a mini presentation on one of the methods described on page 157 of the Student's Book. Their presentations can be either electronic or verbal and they should use the information in the Student's Book and the internet, if available. Also ask students to name as many ISPs as possible. Ask them to consider the difference between an ISP providing for a family and for a business. Make sure to remind students to consider wireless ways of accessing the internet through smartphones, cellular networks and Wi-Fi hotspots.

Student task: Ask students to present their connection method to the class and discuss.

Discussion: The internet and networks. Show **PPT4_3a** (Slide 1) and discuss the reasons for proxy servers. Explain how they are used in business and especially in schools, where they are used to control content. Show Slide 2 of **PPT4_3a** and cover the use of firewalls, their purpose, and where they are used in home and business.

Discussion: Intranets. Students will best understand this as a website that is only accessible by users within an organisation. Intranets normally run in a browser, just like a normal site, but often contain confidential information. Some organisations will allow staff to access their intranet from outside the company using a username and password. These are called 'extranets'.

Discuss the advantages and disadvantages, as outlined on page 158 of the Student's Book. You could also cover Virtual Private Networks (VPNs) here, describing how they use public networks like the internet to connect private networks.

Student task: **Activity 2** on page 158 of the Student's Book – this could be a piece of written work, a verbal presentation or used as homework.

Give extra support by carefully choosing group work members with a mix of abilities who will support each other.

Give extra challenge by adding additional websites to **WS4_3a**. You could also develop Activity 2 by including other organisations (for example, banks, hospitals and retail) and asking how they might use intranets.

Plenary suggestions

Discussion: Look at the table named **Internet, Intranet or Extranet?** (Student's Book page 158) and write those three headings on the whiteboard. Ask students to come and write a possible scenario under each and aim to create as many as possible. Add real-life examples as far as possible.

You could also discuss the following questions:

- What does the future hold for the internet?
- How much information should schools allow to be online?
- What are the advantages and disadvantages of student-based websites?
- Will websites always have the sort of addresses they have now, and how might we access them in the future?

Assessment suggestions	Activities 1 and 2 could be done as homework or collected in during the lesson for assessment.

4.4 Computer networks in business environments (1)

Learning outcomes
- Develop an understanding of applications of computer networks in:
  - finance departments (such as billing systems, stock control and payroll)
  - school management systems (including registration, records and reports)
  - booking systems (such as those in the travel industry, the theatre and cinemas).
- Associate these examples with batch, online and real-time processing.

Differentiated learning outcomes
- **All students must** be able to identify and discuss the basics of school management systems, computer aided learning and booking systems.
- **Most students should** be able to describe applications for school management systems, computer aided learning and booking systems and identify appropriate uses for them.
- **Some students could** explain applications for school management systems, computer aided learning and booking systems and describe appropriate uses of them.

Resources
- **Student's Book: pages 159–160**
- **Files:**
  PPT4_4a.pptx
  WS4_4a.docx
  WS4_4b.docx
  WS4_4b_Answers.docx

Starter suggestions

Give an outline of the lesson: how networks are used in the real world, specifically in school management systems, computer aided learning and booking systems.

Discussion: Like any large organisation, schools contain a number of network systems. With students' help, create a mind-map on the whiteboard of each type of processing with real examples and refer to this throughout the lesson. This lesson will focus on school-based systems before moving to online booking systems, another system student's may be familiar with.

Main lesson activities

Use **PPT4_4a** and talk though the two key networks within Excelsior Academy.

Discussion: How many of these systems are recognisable in your own school? Hand out **WS4_2a** so students can add any explanations or additional notes of any terms they don't understand.

Student task: Ask students to carry out **Activity 1** on page 159 of the Student's Book. They can use **WS4_2a** to do this. Discuss answers and make sure all students understand the various elements of the network. Students should return to the teaching and learning network after the next discussion, alternatively these can be discussed together.

Discussion: Move onto Excelsior's' teaching and learning network, elements of this should be familiar to students. What elements already exist in your school and what advantages do they offer? Students can return to **WS4_4a** is required.

Student task: Ask students to complete **Activity 2** on page 159 of the Student's Book and discuss students' suggestions. Point out that some school networks are allowing limited student and parental access, which means they can view recent results and attendance data away from school. What are the advantages and disadvantages of this technology?

Discussion: Move on to online booking systems. Talk through the example of the cinema booking system given on page 160 of the Student's Book. Ensure each step is clear to students before moving on to the next.

Student task: Create a short, hand-drawn comic book, or flowchart diagram if preferred, showing the process of a person buying tickets. Students should make sure their diagram explains the process clearly. If work can be projected to the class, ask for volunteers to explain their work or ask students to swap their work for peer assessment. Discuss students' work, ensuring all the stages are understood.

> Give extra support by providing a partly completed answer to **Activity 3**.
>
> Give extra challenge by asking students to investigate additional online booking systems like those mentioned on page 160, for example, airport systems.

Plenary suggestions

Discussion: What other ways can online bookings be used and what are students' experiences of them?

How will these systems develop in the future? Especially in schools?

Activity 3 (Student's Book page 160) would be ideal as homework. Hand out **WS4_4b** and collect in the completed worksheets at the next lesson for assessment. Make sure time is given to discuss the answers.

Discuss additional network examples that use batch and real-time processing.

Assessment suggestions	Activity 3 could be given as homework and collected in.

4.5 Computer networks in business environments (2)

Learning outcomes

- Develop an understanding of applications of computer networks in:
  - banking (including Electronic Funds Transfer (EFT), ATMs for cash withdrawals and bill paying, credit/debit cards, cheque clearing, phone banking, internet banking)
  - medicine (including doctors' information systems, hospital and pharmacy records, monitoring and expert systems for diagnosis)
  - libraries (such as records of books and borrowers and the issue of books)
  - expert systems (for example in mineral prospecting, car engine fault diagnosis, medical diagnosis, chess games).
- Understand the importance of confidentiality and security of data in all these applications, including problems surrounding common network environments.

Differentiated learning outcomes

- **All students must** be able to identify and discuss the basics of banking, medicine and expert systems.
- **Most students should** be able to describe applications of computer networks for banking, medicine and expert systems, and identify appropriate uses.
- **Some students could** explain applications of computer networks for banking, medicine and expert systems, and describe appropriate uses of them whilst considering issues like confidentiality.

Resources

- **Student's Book: pages 161–164**
- **Files:**
  PPT4_5a.pptx
  PPT4_5b.pptx
  WS4_5a.docx

Starter suggestions

Give an outline of the lesson: how networks are used in the real world, specifically in banking, medicine and expert systems.

Discussion: How do students imagine these organisations will use networks? These responses could be written down on the whiteboard and referred back to in the plenary.

Main lesson activities

Discussion: The first area covered is the use of ICT in banking, outlined on page 161 of the Student's Book. This is an important part of the syllabus and referred to in many units.

Student task: Ask students to turn to the table on Student's Book page 161, showing the process of using an ATM, into a yes/no flowchart. Also mention how banks use intranets for internal information (this will be developed further in later sessions).

Discussion: Move on to medicine (Student's Book page 162). Show **PPT4_5a** and talk through the process of how networks are used. **Activity 1** could be done on paper or simply discussed as a group.

Discussion: Having discussed patient records and types of data they hold, move onto pharmacy records. As before, discuss the importance of security and privacy in any medical related data. Expand this if possible to the advantages and disadvantages of storing this information online, allowing patients to be accurately treated anywhere.

Student task: **Activity 2** (page 162). Students would benefit from internet access for this activity in order to research jobs within a medical centre.

Student task: **Activity 3** (page 162). Again, this could be written or done as part of a Q&A session in class.

Discussion: Move on to expert systems (Student's Book page 163). Explain each of the key terms and discuss the advantages and disadvantages of such a system. Using **PPT4_5b** as a starting point, either expand on the whiteboard or ask students to expand on paper the mind map to cover the key expert system elements. How many other scenarios can students come up with that would use such systems?

Student task: Ask students to complete **Activity 4** using **Worksheet WS4_5a** and discuss their answers when complete. Not everyone will have the same answer – some tasks could be argued either way, as different garages may do things in different ways.

> Give extra support by adding more information to **WS4_5a** before handing it out, depending on how independent students are and how willing to add notes from the Student's Book.
>
> Give extra challenge by asking students to add explanations to their worksheet answers.

Plenary suggestions

Discussion: What would change if the methods in this session were transferred to other industries outside of those discussed so far. How might these systems develop in the future? Compare the theory covered to the student ideas in the starter.

Assessment suggestions	Any of Activities 1 to 4 could be collected in for formal assessment.
	Activity 4 could be peer-assessed by swapping papers.

4.6 How does ICT help business and personal communication? (double lesson)

Learning outcomes
- Show an understanding of interactive communication applications (such as blogs, wikis and social networking websites).
- Describe and explain the use of communication applications (such as the internet, email, fax, web, video and audio conferencing, cell phones and internet telephony services).

Differentiated learning outcomes
- **All students must** be able to identify and discuss modern communication methods and technology.
- **Most students should** be able to describe modern communication methods and technology including email, SMS, video conferencing, blogs, social networks, and mobile telephone applications and describe business uses of each.
- **Some students could** be able to explain modern business uses for communication methods and technology including email, SMS, video, web and audio conferencing, blogs, social networks, and mobile telephone applications.

Resources
- **Student's Book:** pages 165–168
- **Files:**
  PPT4_6a.pptx
  WS4_6a.docx
  WS4_6b.docx

Starter suggestions

Discussion: Outline the lesson and show **PPT4_6a** to map out the main topics for this lesson and discuss with students their experiences of each one. Write their comments around the mind-map, if possible.

Main lesson activities

Discussion: Email. Students will be familiar with email, but how is it used in business? Describe the difference between web-based email and using an email program like Outlook on a local computer. Use the whiteboard to demonstrate the structure of an email address, including the name, domain, location, etc.

Student task: Hand out **WS4_6a** and ask students to complete **Activity 1** (Student's Book page 166). This could be set as a homework activity if time is short and students have access to the internet away from the classroom.

Discussion: Move on to SMS and explain the possible business uses outside social use. For example, messages and decisions can be made instantly rather than waiting for an email. SMS is more useful for quick communications.

Discussion: Move on to video conferencing. This technology has been around for years but is becoming more commonplace. Use Apple Facetime, Zoom and Microsoft Teams as examples. Ask students to identify the advantages and disadvantages of video as a business tool. Answers could be written on the whiteboard. Follow up with any points missed from the Student's Book. Also explain the technology required by both participants. Touch upon working remotely and sharing information. Follow up with audio conferencing, much more cost effective than a traditional conference call. Discuss the uses for groups of people being able to join the same conversation at once.

Discussion: Web conferencing. Introduce this topic as a successor of both video and audio conferencing in how it allows the combination any online multimedia element.

Class activity: If possible, set up a video or audio conference between one classroom and another, or even a different school if possible.

The first part of the double lesson ends with **Activity 2**.

If this session is being delivered as two single lessons, recap as required.

Student task: Ask students to tackle **Activity 2**, using **WS4_6b** and focusing on the technology covered in the lesson. The internet could also be used for research. Discuss the results with the class. Again, this could be set as homework if time is short.

Discussion: To start the second part of the lesson, move on to blogs, wikis and tweets. Again, students will be familiar with these technologies for personal use, but here you should focus on the business uses. Many large organisations build blogs and Twitter into their own sites – Demonstrate with a range of world news websites, many of which allow comments from readers following the story.

Student task: Ask students to continue filling in **WS4_6b**, adding the advantages and disadvantages of blogs, wikis and tweets.

Discussion: Move on to social networks. Again, these will be familiar to all students, but businesses are now using them more and more. If possible, show commercial Facebook sites, such as New York Times, Microsoft or Coca Cola. Discuss the advantages and disadvantages and how companies are using social network sites to research possible job applicants before interview.

Student task: Set **Activity 3** (Student's Book page 168).

Discussion: Move on to telephone applications. Make sure that students understand the difference between landline, mobile and internet or VoIP systems. Internet-based systems are popular for long-distance calls. These technologies now intertwine with video conferencing as mobile internet speeds reach almost landline speeds.

Student task: Set **Activity 4** (Student's Book page 168). Again, this can be written into **WS4_6b** or done as a separate piece of work.

> Give extra support by allowing Activity 1 to become a small group activity.
>
> Give extra challenge by asking students to complete Activity 3 as an electronic presentation and ask them to present their thoughts to the class or compare and contrast as many of the methods in activity 2 to each other.

Plenary suggestions

Discussion: How might the systems discussed in this session develop in the future?

The results from **WS4_6b** could be talked though or assessed.

How might web, video and audio conferencing change the way we do business and communicate internationally for both commercial and family life?

Assessment suggestions	One or more of Activities 1 to 4 could be done as homework. **WS4_6b** could be collected for assessment.

4.7 Keeping computer network data confidential and secure

Learning outcomes

- Understand the terms user id and password, stating their purpose and use.
- Describe the importance of confidentiality and the security of data, including problems surrounding common network environments.
- Explain the need for encryption and authentication techniques (including the use of user identification and passwords) when using common network environments such as the internet.

Differentiated learning outcomes

- **All students must** be able to describe the importance of keeping data secure with examples of where and why.
- **Most students should** be able to explain the importance of keeping certain data secure and confidential, some of the processes involved and the types of organisations that use these processes.
- **Some students could** explain modern data encryption methods, the reasons for their use and appropriate real-world examples.

Resources

- Student's Book: pages 169–171

Starter suggestions

Discussion: Usernames and passwords. Ask students how many of them use pets, friends, family names and even the word 'password' when creating login details for websites. If someone were trying to guess their details, this would be the best place to start. Look up the top 10 most used passwords online and show to students if possible. Outline the objectives of the session.

Discussion: Write on the whiteboard the following headings: Teacher Access and Student Access. Around them add the information held in the school that students think they should have access to. Make sure students explain their answers and make suggestions such as "Should I be able to look at the salary of the head teacher and should you be able to read the medical notes of a teacher or fellow student?". Add any types of information that are missing, highlighting personal and secure information.

Main lesson activities

Outline the lesson: confidential information and measures to protect it. Starting with student logins if they have them, discuss the need for usernames and passwords. A username is essential for identification and usually does not need to be secure; it may often be a combination of a name and reference number. Refer back to the comments about passwords at the beginning of the lesson and demonstrate how a more secure password can be created by following the password advice listed in the Student's Book.

Student task: Ask students to complete **Activities 1** and **2** on page 169 of the Student's Book. There are password checking websites you may show to students but ensure no real passwords are entered and explain the security risk involved in using them.

Discussion: Data confidentiality. Personal, medical, academic and banking data are all examples we would like to keep secret, but many organisations have access to it. Shop loyalty cards for example hold information about everything we buy, and online shopping sites store our banking information. Discuss how and why this is kept secure.

Student task: "What would be the benefit to someone and danger to you of recording what you do on a computer and the sites you access?" Discuss their answers and point out that this is the basis of spyware.

Discussion: Present to students the UK Data Protection Act and point out the most governments have a list of data protection principles designed to protect the rights of people and how their information is used.

Student task: Set **Activity 3** on page 170 of the Student's Book. This could be written down or done as a class discussion; for example, what might happen if personal data (medical, academic, etc.) were accessed by the wrong person.

Discussion: Encryption. A basic description of encryption is provided in the Student's Book (page 170). The internet could also be used to find a fuller explanation.

Discuss as many real examples as possible, including those in the **Real World** panels on page 170. Which sorts of businesses will rely on encryption of data?

Student task: Set **Activity 4** on page 171. This could be written in class or done as homework.

Discussion: Authentication. Discuss the three factors listed in the Student's Book on page 171 and then the examples in the **Real World** boxes. Expand this to the introduction of biometric access systems now being embedded in smartphones and household devices.

> **Give extra support** by making sure you circulate among students, identifying and remedying problems as required.
>
> **Give extra challenge** by developing Activity 4 into an electronic presentation to present to the class.

Plenary suggestions

Discussion: How might data be kept more secure in the future?

Search for articles and information relating to, for example, quantum-safe encryption.

| Assessment suggestions | Students' responses to Activities 1 to 4 could be collected for assessment, if carried out as written activities. |

4.8 An introduction to website design

Learning outcomes
- Students should show an understanding of the different methods available to design and create websites.
- To understand and implement in later sessions the concept of web design development layers.

Differentiated learning outcomes
- **All students must** be able to describe the different ways in which a webpage can be created and the importance of web design development layers.
- **Most students should** be able to explain the importance of choosing the most appropriate web design tool and describe each of the development layers in detail.
- **Some students could** explain the key differences between open source and commercial web design routes.

Resources
- **Student's Book: pages 174–175**
- **Files:**
  PPT4_8a.pptx
  WS4_8a.docx
  WS4_8b.docx

Starter suggestions

Discussion: On the whiteboard, write the heading: "How are websites made?" and invite students to provide as many ideas as possible in the form of a group mind-map. If any students have friends and family in the profession, or create websites themselves, then ask them for real examples. Highlight, using three different colours, any responses that fall into the three methods outlined on page 174 (HTML, web design software or website CMS) and leave on the board to come back to later in the lesson.

Main lesson activities

Outline the lesson: Looking at the alternative methods of creating website content and some of the concepts behind web design.

Student task: Divide the class into three groups and give each class a copy of **WS4_8a** to complete. Set a timer for 15 minutes and during that time the students must research and create a short verbal presentation based on their worksheet notes. Only two students can present their findings to the class and students could decide themselves on the most appropriate students for the task. Once the time is up, students should present their findings, highlighting any elements missed or not clearly defined. Once complete, show examples of all three methods, such as the ones highlighted on page 174 of the student guide and point out that during the practical element of this unit that the focus will be on programming in HTML. Once complete, refer back to the mind-map created at the start of the lesson, how have the students developed their initial thoughts on website design?

Student task: In pairs or small groups, students should research current examples of web design software. They should consider open source and commercial software, the features they offer and the amount of web design knowledge they require to be used. Discuss their answers.

Discussion: Show **PPT4_8a** and talk through the three web design development layers. Point out to students that this is a set of principles used by many designers but it is not the only one. With a little research on the internet it is easy to find many designers saying there are many more layers and some saying they do not refer to them at all! Explain briefly that CSS are sets of rules for content, a little like heading styles in word processing software, and a more detailed analysis will be made in later sessions.

Student task: Ask students to carry of **Activity 2** using **WS4_8b**. The first example has been created for them and they should then analyse further webpages.

> Give extra support by using group work as much as possible. Make sure to explain the example given in **WS4_8b** to ensure understanding.
>
> Give extra challenge by asking students to do a more detailed comparison of a leading open-source and commercial web design package. Students could also investigate how the concept of web design development layers is considered online.

Plenary suggestions

Discussion: What do students think might be the most appropriate methods for the following people if they wanted to design a website:

- students
- small local businesses
- large international businesses
- older people
- website designers?

Assessment suggestions	Activity 2 could be extended as homework and assessed.
	Written work: Ask students to consider what the future may hold for creating websites.

4.9 Using HTML to create and edit webpages

Learning outcomes
- Students should be able to create webpage(s) including features like setting the background, text colours and formatting using HTML.

Differentiated learning outcomes	Resources
• **All students must be able to create a simple webpage using HTML coding and recognise basic tags to change the page's appearance.** • **Most students should be able to create a webpage using HTML, use tags and attributes to change the page's appearance.** • **Some students could create a webpage using HTML coding and display understanding of how tags, attributes and nesting are used to change the appearance of the page.**	• **Student's Book: pages 176–182**

Starter suggestions

Discussion: Show an example of a popular website on screen and within the browser change the view to 'source view' to show the HTML script. Outline the practical element of this unit: writing and creating a website. Point out to students that commercial sites are rarely written in HTML from scratch and theirs will not be as complex. Ask students to try and identify any elements on the script and suggest what they may do. Outline the project they will undertake: creating a website for a computer supply company (as outlined on page 176 of the Student's Book).

Main lesson activities

Discussion: Make it clear to students the importance of file management and organisation when they work on websites, as links can easily be lost. It would be wise for students to create a new empty folder in their document area for the practical work in this unit.

Please note: The web design sessions in the remainder of this unit require a number of resource files. These need to be accessed during lessons, ideally made available on the school network. Students should be able to copy them to their own user area for working on in class.

The use of large teacher-led demonstrations on the whiteboard is advisable throughout this session.

Many of the demonstrations could be grouped together or even followed independently by students. You can vary this, depending on the group.

Many students will prefer the approach of you demonstrating a single step, students repeating, you demonstrating the next step, and so on.

Demonstration: Using a simple text editor such as Notepad, create the example index page shown in the Student's Book (page 176). It would be wise to create this with students from scratch so they can understand the symbols and syntax used. Show how to save as a HTML file and discuss the importance of file extensions in web design. Depending on the text editor, it may be able to save and edit work in .html format, alternatively it may be required to save work as .txt files. Another issue to point out is the similarity of the letter (l) and the number (1).

Student task: Students should now create the same example as shown; index.html. Remind students that this should be saved in their web design folder.

Demonstration: Make the changes as demonstrated as shown on page 177 of the Student's Book and save it appropriately.

Ask students to complete **Activity 1** on page 178, working in pairs. Ask students to help each other if necessary. Monitor the group carefully and offer support when required. Coding can be frustrating when unexpected errors arise.

Demonstration: Show how to add bold, italic and colour tags to the text in the file. Make sure you use the text version of the file. Discuss the internet's use of RGB and the standard set of web colours available.

Student task: Ask students to complete **Activities 2** and **3** (page 178). Again, ask students to help each other and monitor the group for anyone getting behind.

Demonstration: Focus on text and page formatting, with a reminder of how to preview work in a browser. Demonstrate how to print from text editor and browser.

Student task: Set **Activity 4** on page 179 of the Student's Book. Students could also annotate their coding, identifying important elements to show their understanding. The answers to question 4 could be written on the reverse of their printouts.

> **Give extra support by**:
> - Allowing students to work in pairs and guide students
> - Study the **Code Academy** HTML glossary in their own time:
> - Experiment with live HTML scripting using a free HTML editing tool
>
> **Give extra challenge** by asking students to do the following:
> - Demonstrate to their peers in small groups or in front of the class.
> - Go to the **W3Schools** website and encourage them to experiment with different tags and coding elements.
> - Go to the **Tutorials Point** website and complete the **Introduction to HTML** exercises. These would also make great independent homework activities.

Plenary suggestions

If time allows, students could create a new simple page with content of their own choosing but using the skills learnt.

Hold a Q&A session: Ask students to describe a selection of the keywords used in the lesson.

Show again the professional websites from the starter and show the coding. How much more information can they understand now than at the start of the session?

Assessment suggestions	Answers to Activity 4 could be collected and assessed. A student-designed webpage, on a topic of their choice, could be an assessed homework activity.

4.10 Creating and using cascading stylesheets

Learning outcomes
- Students should be able to create and apply cascading style sheets to HTML documents.

Differentiated learning outcomes
- **All students must** be able to describe the use of cascading style sheets and apply a sample.
- **Most students should** be able to explain the importance of creating relevant style sheets and create their own.
- **Some students could** create and apply stylesheets and experiment with style sheet hierarchy.

Resources
- **Student's Book: pages 183–186**
- **Files:**
  PPT4_10a.pptx
  PPT4_10b.pptx

Starter suggestions

Discussion: Show students the slide **PPT4_10a** and discuss an example three-page website like that in the Student's Book. Each page can be edited individually to maintain a corporate house style but a stylesheet allows the same elements across three, or three hundred, pages to be set simultaneously. Also remind students how important organising their files and folders is essential to prevent 'broken links'. If possible, show an example of a broken link. Remind students that this forms part of the presentation layer as described in the web development layers.

Main lesson activities

Student task: Carry out **Activity 1**, creating a folder, sub-folders and copying their index.html file into website folder just created.

Demonstration: Show students the CSS file from the example website and demonstrate how by changing one aspect, body text for example, that the pages it links to are instantly changed when previewed in a browser. Demonstrate how using a text editor a CSS file can be created and saved in the correct format. In order to use a style sheet, existing inline styling should be removed.

Student task: Complete **Activity 2** as shown, editing their index.html file.

Demonstration: Show students how to create a new CSS file and specify the rules within it. Make sure students understand the formatting of new rules (full stops, no spaces) and point out the limited number of fonts. The choice of fonts and colours should obviously reflect the house style of the organisation and students should be reminded of this. The agreed choice of fonts in-built into most website software is to ensure consistency and usability of webpages anywhere in the world. If a designer chooses an obscure font that a user does not have, then problems can appear when viewing the site in another browser, hence a limited, recognised selection.

Student task: Carry out **Activity 3**, creating a CSS file, similar to the one shown in the Student's Book and attaching it the index.html page. Extra support can be given by provided partially completed CSS files for students to edit rather than create. They can experiment with fonts, formatting and colour.

Demonstration: The use of six heading tags is common across web design and these can be specified within a CSS file. Demonstrate using the ACSSupplies file to create a similar h1 rule.

Student task: Carry out **Activity 4**, creating an h1 rule and setting formatting.

> Give extra support by pointing students to online tutorials on the **W3Schools** website. These could be done independently for homework.
>
> Give extra challenge by allowing students to create additional heading styles in Activity 4, and if they started their own website in the last session they can add CSS styling to it.

Plenary suggestions

Discussion: With so many rules both within webpages and attached style sheets sometimes it difficult to know which settings will overwrite another. **Show PPT4_10b** and explain how the browsers decide on style sheet hierarchy. Point out that the diagram is not an exact science, rather just a guide that can be used when creating websites. Inconsistencies can appear

Assessment suggestions	Make sure to use formative assessment throughout the session, monitoring students' progress through the activities and asking for screen prints at certain stages.
	An additional project – for example, designing a new school page or a page based around a club or hobby – could be run alongside this unit and taken in for assessment.

4.11 Using HTML to add website content
(double lesson)

Learning outcomes
- Create webpage(s) including features like: menu options, text and graphic hyperlinks, setting the foreground, background and text colours.
- Use tables to organise a webpage.
- Insert an image in a webpage and place the image relative to text and other objects.
- Carry out basic image manipulation, making the image suitable for internet use.

Differentiated learning outcomes
- **All students must** be able to create webpages using HTML, use tables and images to change the appearance and create links between pages.
- **Most students should** be able to create, open and edit a webpage using HTML, use and format stylesheets, tables and images to improve its appearance and add hyperlinks for navigation.
- **Some students could** create, open and edit webpages using HTML, improve the appearance using images, tables and cascading stylesheets, use advanced image and website editing features and add hyperlinks for internal and external links.

Resources
- **Student's Book: pages 187–197**
- **Files:**
  PPT4_11a

Starter suggestions

Outline the lesson: extending the website, adding more structure to webpages using tables, and adding hyperlinks, images, sound and video. Display **PPT4_11a**, the diagram provided by the computer company – this is a plan of the pages and resources needed. If possible, hand out copies of the slide to students so that they can annotate it as required or tick elements off when complete.

> **Please note:** The images and key elements needed for this session are included in both the student's and teacher resources. Ideally, students should already have access to them in their local document folders. However, students should also be free to create or find their own elements and this should be encouraged. Basic image manipulation is covered in this session, but you may need to allow additional time if students wish to develop skills in this area.

Main lesson activities

Demonstration: Show how to create a new file (saving it as Page2.html), adding a table and image using the examples as shown in the Student's Book and reminding students about the importance of file management and locations when using hyperlinks. Depending on the group, this could be divided into separate demonstrations for tables and images. When demonstrating images, make sure students are aware of the location of the cursor when inserting images.

Student task: Ask students to carry out **Activity 1** (Student's Book page 193) using the Student's Book digital download file where indicated. If students are planning to use their own images, they should wait until after the next demonstration.

Discussion: Using image-editing software to resize and format an image for web design use. The example shown in the Student's Book uses a free open-source package called GIMP. Explain the image editing terminology and file formats. Also discuss why file format and size are important in web design and the link between size and loading times.

Demonstration: Demonstrate how to download an image, resize it, change the colour depth and save it to a suitable format, such as JPEG. Make sure students are aware of how experimenting with colour depth and compression settings can lower file sizes but reduce the quality of the image.

Student task: Give students the opportunity to experiment with downloading images, resizing and formatting them. When they are happy with these skills, they can carry out **Activities 2**, **3** and **4** on pages 195-196 of the Student's Book.

The first part of the double lesson ends after Activity 4.

Give extra support by asking those with website design experience to support others.

Give extra challenge by showing students how to cut out images within the image editing software save with a transparent background.

Plenary suggestions

Discuss student findings after experimenting with images and allow time for students to complete the activities in this part of the session.

Assessment suggestions	Make sure to use formative assessment throughout the session, building checkpoints into the lesson to monitor progress.
	Ask students to write a short report on the alternative file types available for online images and the advantages and disadvantages of each.

4.11 cont. Using HTML to add website content (double lesson)

Learning outcomes
- Create webpage(s) including features like: menu options, text and graphics hyperlinks, setting the foreground, background and text colours.
- Creating internal and external links and use anchors.
- Use tables to organise a webpage.
- Attach an external stylesheet.
- Inserting and placing appropriately multimedia elements into a webpage.
- Carry out basic image manipulation as required

Differentiated learning outcomes
- **All students must** be able to create webpages using web authoring software, use tables and images to change the appearance and create links between pages.
- **Most students should** be able to create, open and edit a webpage using web authoring software, use and format stylesheets, tables and multimedia objects to improve its appearance and add hyperlinks for navigation.
- **Some students could** create, open and edit webpages using web authoring software, improve the appearance using multimedia, tables and CSS, use advanced image and website editing features and add internal, external hyperlinks and anchors.

Resources
- **Student's Book: pages 187–197**
- **Files:** PPT4_11a

Starter suggestions

If this session is being delivered as two single lessons, a recap will be required.

Please note: For this session, students will again need access to the images and key elements on the Student's Book digital download.

Main lesson activities

Student task: Ask students to carry out **Activity 5**, adding images to their web pages, using the skills they have gained so far. Students may need extra time if they would like to add their own images. Alternatively, these images could have been collected for homework.

Demonstration: Using the resources provided, show students how to add sound and video into the home page (**index.html**) of their website. Remember that any assets they use need to be saved into the media folder of their website.

Student task: Carry out **Activity 6**, adding sound and video to their webpage.

Please note: Depending on the group, an additional lesson for students to simply work on creating extra pages may be required.

Discussion: How hyperlinks work. Students should already be familiar with the term hyperlink from presentation work or adding links to DTP work or spreadsheets. The principle is the same. Also include bookmarks and the difference between internal and external hyperlinks.

Demonstration: Show how to create and add hyperlinks to text and images to every page in order to navigate around the site. Additional link text or graphical buttons can be added as required.

Discussion: Outline the differences between relative and absolute file paths. An absolute file path become irrelevant if moved to another computer. This is why relative file paths are used, keeping within the website directory structure. Make sure students understand the key differences understand the differences between the examples in the book.

Student task: Ask students to tackle **Activity 8** on page 194 of the Student's Book.

Demonstration: Show how to create links to external pages from both text and images and setting the target option. Also discuss how important it is to check links. Including a deliberate mistake could be a good way of showing how to edit and correct a mistake. Demonstrate how to add an email address link and the variety of printing options available from most browsers. also include how to access the 'View code' option within each browser, providing users a chance to view the code behind the sites they view. Point out that large commercial sites will have much more complex HTML code due to the fact they are often 'live' sites, generated using content management systems rather the static site students are creating.

Student task: Ask students to tackle **Activity 9** on page 196 of the Student's Book. Students should make sure they test all links before the session ends.

Demonstration: Discuss and demonstrate the use of <div> tags to create areas of a webpage that can be specifically formatted using direct formatting or using CSS.

Student task: Ask students to format a table like the example shown and discuss how else <div> tags could be used in their own, and larger sites.

> Please note: Again, this may be a point where additional time for students to work through creating their pages may be required.

> Give extra support by providing all images and a CSS file for students to edit and use and asking those with website design experience to support others.
>
> Give extra challenge by allowing students to create their own images, sounds and video for the site, keeping the target audience in mind. Allow students to experiment with embedding external codes from Vimeo or YouTube; these should be pasted into the HTML code.

Plenary suggestions

A common problem is the links from pages to images being lost if any of the files are moved once used in the site, has this happened during the lesson.

Discussion: What other things would students like to be able to add to their website?

Assessment suggestions	Build formative assessment checkpoints into the lesson to monitor progress.
	Print and proofread website in preparation for next lesson.

5.1 The use of microprocessor-controlled devices in transport systems and manufacturing

Learning outcomes
- Be aware of the benefits and drawbacks of the introduction of new technology into the workplace.
- Be able to describe the effects of such changes on the business, their products and the employees.

Differentiated learning outcomes

- **All students must** be able to state an advantage and a disadvantage of the introduction of technological change and of home-based working.
- **Most students should** be able to recall a number of advantages and disadvantages of the introduction of new technology and home-based working and attempt to draw conclusions on the basis of these advantages and disadvantages.
- **Some students could** compare the suitability of different businesses to the incorporation of new technology and home-based working.

Resources
- **Student's Book: pages 202–204**
- **Files:**
  WS5_1a.docx
  WS5_1b.docx
  WS5_1c.docx
  WS5_1d.docx

Starter suggestions

Discussion: The changes due to the introduction of technology.

Ask the class: What effect will the introduction of technology have on transport and manufacturing? How will it affect the way the businesses operate? Use the examples on page 210 of the Student's Book as prompts.

Consider wider issues: Are the changes always beneficial? If so, which groups in society will gain advantages and which groups will suffer disadvantages?

Consider the effect nationally and globally of introducing technology: is the introduction of newer technology inevitable? If so, how can a nation manage the changes?

Collect answers on a whiteboard so that they are available for his and future lessons.

Main lesson activities

Discussion: Read and discuss the changing face of transport cited in the Student's Book on page 203. Add any new points raised to the listing on the whiteboard.

Student task: Hand out **WS5_1a** and ask students to complete **Activity 1**.

Discussion: Read the section on 'Transport safety' on page 203 and the level of risk we might be happy to take in the future with automated systems. Expand the discussion to include the security of the data these systems will generate.

Student task: Ask students to complete **Activity 2** on page 203, using **WS5_1b**.

Discussion: Consider the issues around the increased use of technology in manufacturing and production control. Explain the increased use of automation and robotics and its relative advantages and disadvantages. Ask the class: What products or services are you and your family using that include automation? Pick out other criteria that may present the students with difficulty in answering **Activity 3**, such as 'consistency'. You could give the example of 'Build quality', particularly in relation to cars; explain the advantages to the consumer of cars that have been built to a consistent, high standard.

Student task: Ask students to complete **Activity 3** (Student's Book page 204), using **WS5_1c**.

Discussion: Work through the paragraph 'Humans verses computer control' on page 204. Consider the impact on society of human interaction with new technology, from the need for new training and skills to the loss of jobs and the relationship we have with technology.

Student task: Ask students to complete **Activity 4** (Student's Book page 204), using **Worksheet 5.4**.

> Give extra support by inviting students to imagine that they are working from home. What changes would be made to their working conditions, productivity and social life? What effect would it have on their home? Would changes have to be made?
>
> Give extra challenge by considering what jobs are suitable for working from home. What jobs cannot be done in this way?

Plenary suggestions

Review the lesson. Question the class about their preferences for working either from home or in a workplace and ask them to give their reasons. Have any students changed their mind over the course of the lesson?

Discuss the implementation of new technology:

Do firms have any choice (i.e. is it inevitable that new technology is introduced)?

Will all firms need to implement technological change?

What areas of work would be ideal for the introduction of new technology?

Conversely, which businesses would be advised to avoid technological change?

Ask the students to summarise the plenary.

Assessment suggestions	The students' own summaries of the plenary could be the basis for an assessment.

5.2 The impact of satellite systems

Learning outcomes
- Be able to describe the characteristics of GPS and GIS.
- Explain how GPS and GIS can be used in navigation and communication.
- Describe the advantages and disadvantages of GPS and GIS.

Differentiated learning outcomes
- **All students must** be able to identify examples of GPS and GIS systems.
- **Most students should** be able to discuss ways GPS and GIS can be used in navigation and communication.
- **Some students could** compare the advantages and disadvantages of GPS and GIS.

Resources
- **Student's Book:** pages 205–206

Starter suggestions

Discussion: What impact do satellite-based systems have on our day to day lives?

- Ask the class: What effect has instant worldwide communication had on the way we work and socialise?
- Consider wider issues: Are the changes always beneficial? If so, which groups in society will gain advantages and which groups will suffer disadvantages?

Collect answers on a whiteboard so that they are available for later exercises.

Main lesson activities

Discussion: Read and discuss the section **Global Positioning Systems (GPS)** on page 205 and discuss the basic technological concepts behind satellite communication.

Student task: Students to complete **Activity 1** on page 205 of the Student's Book.

Discussion: How do the answers from **Activity 1** group together? Can GPS-based devices be categorised?

Discussion: Read the section on the use of Geographic Information Systems (GIS) and how they embed exact location data with additional information relevant to that location. This might include population, environmental or governmental data.

Student task: complete **Activity 2**, Investigating the many satellite platforms in use around our planet and the many more proposed. Additional research time may be required. This could also be a homework project.

Discussion: Read the remainder of the text in the session, which focusses on the satellite navigation, communication and the advantages and disadvantages of these systems. Do students agree with all of the statements and what additional examples can they offer?

> Give extra support by inviting students to consider a number of scenarios, school or employment based, and consider the use of technology discussed in this session. Examples might include being able to provide an exact location if lost or making cell phone calls from any location on earth.
>
> Give extra challenge by asking students to consider the future of satellite technology and the possibilities for expansion in already crowded skies.

Plenary suggestions

Review the lesson. Ask students to link the ideas from this lesson with those of the previous session where teleworking was discussed.

- Do firms have any choice but to offer more flexible working arrangements?
- Will all firms take up the challenge of offering employees more freedom over working hours?
- What areas of work would be ideal for the introduction flexible working?
- Conversely, which businesses could not operate with the introduction of flexible working?

Ask the students to summarise the plenary.

Assessment suggestions	The students' own summary of the plenary could be the basis for an assessment.

5.3 Health issues

Learning outcomes
- Be aware of the possible risk to health when using computers for extended periods of time.
- Develop strategies to be taken to minimise the risk to health when using computer equipment.

Differentiated learning outcomes
- **All students must** be aware of health problems that can arise as a result of regular use of computers.
- **Most students should** be able to suggest ways of reducing the risk of developing discomfort and possible injuries as a result of the regular use of computers.
- **Some students could** recognise that perfect practice is rare and will involve a wider range of considerations than setting up equipment at the workstation.

Resources
- **Student's Book: pages 207–208**
- **Files:**
  WS5_3a

Starter suggestions

Discussion: Ask the class to examine how everyone is sitting:

- Could their posture be improved?
- Why should it be improved?
- What problems do people know about that are associated with bad posture?
- Why is this a particular problem for those who regularly use computers?
- Has anyone heard of RSI? What is RSI?

Discuss what can cause RSI and what signs need to be looked out for.

Main lesson activities

The topic of **Health issues** is very commonly considered in connection with **Safety issues** which is covered in Session 8.1 (Student's Book page 340), so it might be an idea to go on to that session after completing this one.

Discussion: Review the problems that might occur because of bad posture when using a computer. Define the term repetitive strain injury (RSI).

Student task: Ask students to work through **Activity 1** (Student's Book page 207), using **WS5_3a** for part 3.

Discussion: Read through the table on **Minimising the risks to health** on page 207 of the Student's Book. As you work through each of the bullet points, consider each device that you have on the computer desk at school and at home. Do any of the points made in the text cause you to think about changing where something is positioned? The space you give yourself? The comfortable distances needed? How are the computer desks/benches at school helping you to be comfortable? Could you redesign them to make them more 'ergonomic'? Are problems you identify due to the arrangement of the devices or are the devices themselves the problem?

Student task: **Activity 2** will help to draw all the considerations of this session together – and should provide useful practical information for all the computer users.

> Give extra support by demonstrating bad posture as well as good posture. Arrange the devices on the desktop in particularly bad positions.
>
> Give extra challenge by asking students to design an 'ergonomic' computer workspace that shows good practice.

Plenary suggestions

Review the health issues covered in the session. Ask students to evaluate their own practice:

- Is it possible to improve their health when using computers for a prolonged period?
- Are there other issues, such as cost, that will affect the creation of an improved workspace?

Assessment suggestions	The design task (mentioned above as an extra challenge) or the posters produced at Activity 2 could be used to assess progress.

5.4 Commercial and ethical considerations
(double lesson)

Learning outcomes
- Understand, in outline, how online shopping, banking, POS and EFTPOS systems operate.
- Recognise that there are advantages and disadvantages of employing these systems.

Differentiated learning outcomes
- **All students must** be able to understand in general terms how online shopping, banking and POS/EFTPOS systems operate, together with the risks and advantages of online shopping and banking.
- **Most students should** be able to discuss the advantages and disadvantages of online shopping and banking.
- **Some students could** consider security systems used in online shopping and banking.

Resources
- **Student's Book: pages 209–212**
- **Files:**
  WS5_4a

Starter suggestions

Discussion: Open the discussion by asking how many students have used their computer to make a purchase in the last week/month. Review what is needed for this type of shopping. Make a list of the items on the whiteboard and ask students to link them to show how the data flows within the system. Leave this on the whiteboard for use later in the lesson.

Main lesson activities

Discussion and student task: Read the text on page 209 of the Student's Book up to the end of the first paragraph under 'Internet shopping'. Then ask students to spend a few minutes on **Activity 1**.

Discussion: Discuss the advantages and disadvantages of online shopping. Review the list on page 209 of the Student's Book and ask students to add more items as suitable.

Student task: Ask students to complete **WS5_4a** in pairs or small groups. Discuss responses in the whole group. Points might include the following:

1. Clothing could be unsuitable as you cannot try the clothes on, although many clothing stores operate as online retailers – often offering a free returns service if the clothes do not fit. They often improve their websites, for example by having good images of the clothes for sale and the facility to zoom in to examine the clothes in more detail. Other unsuitable items might include food and perishable items, medicines and living creatures.

2. Online shopping could have a negative effect on the community, for example, threatening the existence of traditional shops in towns and cities where it is expensive to maintain a high-street shop.

3   Traditional retailers could compensate by offering extra services for items bought from them or a more personal service, for example, bike shops not charging for fitting a new item to a customer's bike.

4   The risks of using online shopping include: the site not being a real shop at all; the customer's card details being stolen and sold to other criminals to create another, false identity for criminal purposes; money being taken from the customer's account and no goods supplied.

5   Security can be improved, for example, using websites employing 'secure sockets layer' of transfer protocols (incorporated into https). Credit card payments might be subject to bank intervention during the authorisation of sale process, for example, by asking for the customer's password or code.

Student task: Ask students to complete **Activity 2** (Student's Book page 209).

Discussion: Move on to discuss POS and EFTPOS systems as outlined on pages 210 of the Student's Book. Draw a diagram on the whiteboard to show how the systems operate; this can be compared with the online shopping diagram drawn earlier.

Discussion: What are the advantages to the store of installing POS/EFTPOS systems? Outline the way the 'just-in-time' stock control system operates. What are the advantages and disadvantages to the store of maintaining such a system? Points to cover include the fact that warehouses do not store so much food; food now moves between outlets in lorries, which has an impact on the road communication network.

Student task: Students could compare the advantages and disadvantages of POS/EFTPOS – to the customer, the store and the community.

Discussion: The use of the internet in providing banking facilities (Student's Book page 210). Point out that some banking activities are not suited to the use of online services, for example, many small-volume local traders are still paid in cash, which cannot be transferred to the trader's account via the internet. Bank staff are needed by their customers for financial advice - talking with an unseen and unknown person over the internet is unsatisfactory.

Student task: Ask students to work through **Activity 3** (Student's Book page 210) in pairs or small groups. Feedback students' answers in the whole group.

Plenary suggestions

End the first part of the lesson by thinking about future developments.
- What is the future of online shopping and banking?
- Will there be an unstoppable move towards more and more online provision?
- What is the future for the traditional high-street shop?

5.4 cont. Commercial and ethical considerations (double lesson)

Learning outcomes
- Recognise that applying ICT solutions have consequences beyond providing a more cost-effective business.
- Recognise that moral and ethical criteria are material to the notion of unrestricted access to the internet.

Differentiated learning outcomes
- **All students must** be able to recognise the advantages and disadvantages of unrestricted internet access.
- **Most students should** be able to discuss the advantages and disadvantages of unrestricted internet access; identify the implications of applying an ICT solution.
- **Some students could** evaluate the need for control of access to information on the internet; be able to discuss issues surrounding the introduction of an ICT solution.

Resources
- **Student's Book: page 209–212**
- **Files:**
  WS5_4b

Starter suggestions

Start by recapping the theme of this double lesson: commercial and ethical considerations of developments in ICT. The first half concentrated on commercial aspects; in this part of the lesson, we will focus on moral and ethical considerations.

Starter discussion: What do we mean by 'moral' and 'ethical'? Discussion should take in the idea of behaviour that is generally accepted to be right and wrong, or acceptable/unacceptable, within a particular society.

Ask students to read the **Tip** (Student's Book page 211).

Main lesson activities

Discussion: Discuss the moral and ethical aspect of the use of ICT as outlined on page 211 of the Student's Book.

Student task: Using **WS5_4b** ask students to work through the questions in **Activity 4** on page 211, working in small groups.

Depending on the facilities available, it would be useful to display the results of the Activity either as notes around the classroom or as contributions to the whiteboard. Students should be able to view the contributions prior to the plenary session.

Discussion: Discuss the pros and cons of policing the internet (Student's Book page 211. Point out that the internet has a marked effect on the availability of information; individuals can use this information for antisocial purposes (for example, organised crime and illegal activities). However, much good happens as a result of free access to information (for example, repressive governments cannot hide the freedom and liberty available to others in nearby countries). Point out that finding a balance between whether to police or not is difficult and has resulted in the monitoring of internet use in some countries.

Student task: Ask students to work through the questions in **Activity 5** (Student's Book page 211), working in small groups.

Depending on the facilities available it would be useful to display the results of the exercise either as notes around the classroom or as contributions to the whiteboard. Students should be able to view the contributions prior to the plenary session.

> Give extra support by leading students through how internet-based systems work using suitable diagrams.
>
> Give extra challenge by questioning. For example: Is the world improved by computer-based information systems?

Plenary suggestions

Review the lesson. Refer to the posted responses to **Activities 4** and **5** and ask the class for their conclusions about what they have learned during the lesson and read in the posted contributions. Ask the class:

- Who should be considered when a new ICT solution is proposed?
- Do companies have a moral and ethical responsibility beyond their employees?
- What can be done about internet crime?
- Does the advent of the internet create new criminals?

Consider the effect on crime of the internet, in terms of the speed at which information travels and the fact that internet crime is international.

The Assessment activity below could be set as homework to provide summative assessment for this session.

Assessment suggestions	Ask students to prepare an extended prose answer to the following question: 'Is the limitless amount of uncensored information available online really a good thing?'

5.5 The need for copyright

Learning outcomes
- Understand why copyright law exists and identify copyright restrictions on software.
- Recognise the limitations that copyright imposes on the user.
- Identify examples of open-source software.

Differentiated learning outcomes
- **All students must** be able to recognise the role of copyright law.
- **Most students should** be able to give reasons for the use of licences and genuine software; outline methods that software companies employ to prevent software piracy.
- **Some students could** compare the advantages of open source and retail software; discuss methods that software companies use to combat software piracy.

Resources
- **Student's Book: pages 213–215**
- **Further resources: Websites given in the text**

Starter suggestions

Discussion: Introduce the term 'copyright.' Outline its purpose. Questions to class:

- Why is copyright used?
- What would happen if there were no copyright laws?
- Can you give any examples of where copyright is used?
- How do you know if something is copyrighted?

There are lots of images of copyright symbols that can be found via a web search – these can be used as background to the starter.

Five-minute task: Ask students to find a copyright symbol that they like and write a warning that is suitable for copyright to protect an original work (such as an invention, music or image).

Main lesson activities

General point: It would be useful for students to be able to access the web during the lesson. Including some element of research will enable you to identify, from students' answers, those students who are highly able and those who need more support.

Discussion: Introduce the concept of copyright protection extending to software.

Questions to class: Is it acceptable to protect software by copyright law? A follow-up could explore the investment made by software firms (for example, number of hours, the amount of testing, the responsibility of the manufacturer to their customers, the rights of the employees). The Microsoft website gives some useful background (for example, how to check that software is a genuine MS product).

Student task: Ask students to carry out **Activity 1** on page 214 of the Student's Book.

Discussion: Ask students if they are aware of the ways, or what to look out for, that they can be sure that the software that they have, or are buying, is genuine. Could they recognise the markers, labels and other 'identifiers' that companies place on packaging to indicate to buyers that the product is genuine. What software methods can software producers employ? If students have recently bought software, what indicated to them it was a genuine product? Are they aware of non-genuine software on offer locally?

What legal protection do software companies have? Does it work? How is it applied?

Discussion: Introduce the concept of open-source software. Explain how the software is produced and give some examples such as *Open Office* (concentrate on this since it is the focus for **Activity 3**), *GIMP, Audacity and Blender.* Point out that you usually are required to register and accept the terms of the licence – the reason for this being so that you can take part in the development/testing of the software and can be alerted to recognised problems and new developments.

Demonstration: How can you be sure that you are downloading from a reputable source? Demonstrate the use of anti-malware programs to alert the user to possibly dangerous sites and stress the need to check the URL to make sure that you are on a genuine site. (Try typing 'free software like Photoshop' into a search engine). Point out that some programs may be genuine, but others may include a free dose of malware too!

Student task: Ask students to carry out **Activity 2** on page 214 of the Student's Book.

Discussion: Outline the problems of file transfer as an attachment in email. One issue is that the attachment may look like a .doc file but may, in fact, contain an executable file that could be an item of malware (virus, trojan, key logger, etc.). Give an overview of safe working practice to minimise the risks.

Student task: Ask students to carry out **Activity 3** on page 215 of the Student's Book.

> Give extra support by circulating during the Activity sessions and asking stepped questions to lead the student to the intended conclusions.
>
> Give extra challenge by asking students to draw a flowchart which outlines the steps that may be necessary to get some software activated during an installation process.

Plenary suggestions

Review the main vocabulary and concepts introduced during the lesson. Question to the class: If viruses and other malware are so destructive, how can a 'mission critical' system, such as the launch of a missile, be protected?

| Assessment suggestions | The written answers to the Activities could be used to assess progress within the teaching group and identify students who need further support. |

5.6 ICT applications: Modelling

Learning outcomes
- Describe the advantages of computer modelling.
- See how computer modelling can help model real-life situations and support future planning.

Differentiated learning outcomes
- **All students must** be able to define the term 'modelling' as applied to financial situations.
- **Most students should** be able to suggest real-world examples computer modelling.
- **Some students could** suggest situations where modelling can support real-world planning.

Resources
- **Student's Book: pages 216–218**

Starter suggestions

Start a discussion about the raising of money to repair roads (potholes, kerbs, repainting markings and so on) by the introduction of a new 'Road User Tax (RUT)'. How will they decide how much money is needed? How much money will be spent in a month, year? How can they know? How can they allocate money? How much does each road user have to pay? Do motorbike users have to pay as much as someone who drives an articulated lorry? Bring the discussion around to future planning, predicting, and building a model? Why model?

Main lesson activities

Discussion: Work through the sections on personal finance, bridge design and flood management, on page 216 of the Student's Book, ensuring that students understand:

- what computer modelling is
- why real-world models are created
- how models can support future planning
- the advantages and limitations of modelling

Student Task: Ask students to tackle **Activity 1** on page 217 of the Student's Book. Make sure to make the direct link between computer modelling and spreadsheet models in the classroom.

Discussion: Review students' answers then take a look at traffic management and weather forecasting.

Student task: Set **Activity 2**. Students would benefit from time to research a local weather forecast system.

> Give extra support by providing local real-life examples, or website links, for students to consider.
>
> Give extra challenge by asking students to expand their research to examples from anywhere in the world, outside of their experience.

Plenary suggestions

What other real-life modelling situation is modelling of benefit, expand suggestions to earthquake modelling, space travel or even pandemic models.

| **Assessment suggestions** | Provide students with an example outside of those provided and ask them to consider the following: What models will need to be made? What data will need to be collected? Who will use these simulations? Who will benefit from them? |

5.7 Microprocessor-controlled devices in the home

Learning outcomes

- State how, in broad terms, microprocessors are used to control the function of familiar household devices.
- Show an understanding of digital and analogue data types and the need to convert data between analogue and digital.
- Understand the relationship between input, processing and output and be able to differentiate familiar control systems into these separate parts of a control system.

Differentiated learning outcomes

- **All students must** be able to recognise input and output data and devices and processing in general terms for functioning systems of familiar domestic devices.
- **Most students should** be able to create a diagram to show the relations between input, processing and control, and recognise that any processing is reliant upon the inclusion of a microprocessor in the design of the device.
- **Some students could** create suitable diagrams showing an understanding of detailed input, output and processing requirements for a system.

Resources

- **Student's Book: pages 219–221**
- **Files:**
  PPT5_7a.pptx
  PPT5_7b.pptx
  PPT5_7c.pptx
  WS5_7a.docx

Starter suggestions

If possible, try and collect a range of household devices like the ones in the Student's Book. The older the better as a good talking point, 'Guess the year' for example.

Discussion: What household objects contain a microprocessor? You could use **PPT 5.1** to record the answers and complete as a whole class activity or set the student task below.

Student task: Make **PPT5_7a** available to students, who then write in the names of six items that contain a microprocessor. They could insert pictures of the items, but a better idea would be to add more detail (for instance, what the microprocessor does in each item). Some students may find dates for first use of a microprocessor in the object and rank the objects as a timeline.

Main lesson activities

Discussion: The increased functionality of household objects as a result of using a microprocessor (Student's Book page 219). Mention how this increased functionality can enhance our lives, such as recording or pausing live TV programme or the ability of a camera to choose the best exposure settings before taking a picture.

Student task: Ask students to carry out **Activity 1** (Student's Book page 219), thinking in general terms about the advantages of labour-saving devices in the home.

Discussion: Feed back students' ideas about the advantages of these devices.

Student task: Ask students to complete **Activity 2** (Student's Book page 220), using **WS5_7a** to record their answers. Extend the activity to include disadvantages.

Discussion: Introduce the microprocessor control system used in some central heating systems and use **PPT5_7b** to display the diagram from page 220 of the Student's Book. Here you could elaborate the processing part of the system: you could introduce the idea of feedback and use of the IF function (depending on the group).

Spend some time on teaching the difference between analogue signals (for example, sound in which the signal varies depending on the volume of the sound) and digital signals (for example, registration at school where marks for presence are all the same signals). Students should be made aware of the need to convert analogue signals to digital in order that the data can be used by the microprocessor.

Student task: Ask students to complete **Activity 3** (Student's Book page 220), using **PPT5_7c** to record their answers.

Discussion: If students have completed **Activity 3** using the PPT slide, the files can be reviewed by displaying them on the whiteboard. Draw attention to the students' use of language; guidance on this appears on page 220 of the Student's Book.

Review the remaining examples of microprocessor control as found on page 220 of the Student's Book. A further example of embedded web technology is found in some burglar alarms, which if activated by an unauthorised entry, rings your mobile phone with a previously recorded message or rings a security service.

> Give extra support by giving simple examples of processing in control systems, such as in the human body: if you get too hot, sweat is produced to cool you down. In this case, the processing of input sensors recording body temperature is done by part of the brain.
>
> Give extra challenge by asking students to detail processing as indicated above, or by asking students to produce an input/processing/output diagram for an oven that has an internet connection.

Plenary suggestions

Review the lesson and if possible link the use of computer-controlled systems in school, maybe ask a member of the technical support team to come and talk about it. Students could then ask questions supported by their new learning this lesson.

Assessment suggestions	The input/processing/output diagrams mentioned in the lesson plan could be used as assessment opportunities. For some students, this could be extended to include more detail about input devices and how data from these are used in the microprocessor to control output to the actuators of the system. The detail could be extended into programming pseudo code for suitable candidates.

5.8 Data analysis: Spreadsheet modelling skills (double lesson)

Learning outcomes
- Incorporate formulae in spreadsheets.
- Use with understanding absolute and relative cell referencing.
- Alter the data in a spreadsheet so that a required outcome can be modelled.

Differentiated learning outcomes
- **All students must** be able to create a spreadsheet containing formulae and some appropriate formatting.
- **Most students should** be able to use absolute and relative cell referencing as required.
- **Some students could**, with understanding, use appropriate cell referencing and alter key data to meet a required outcome.

Resources
- **Student's Book: pages 224–228**
- **Files:**
  SummerFair.csv

Starter suggestions

Discussion: Spend some time refreshing everyone's recall of spreadsheets. Start by demonstrating how to open a spreadsheet and then, with everyone viewing the open blank sheet, ask a few questions, such as:

- What is special about spreadsheet software?
- What does spreadsheet software allow you to do?

We might expect students to offer suggestions such as 'use formulae', or 'make a model'. Investigate this further by asking: Why are models important? Why not just type in the data and use a calculator to get the answers?

It's a good idea to display the answers, both so that students can see where the questioning is leading and as a reminder for future reference. Depending on the time available and the group, you might find it useful to recall some of the automated tasks available in spreadsheets, such as filling, recalculation, the use of lookup routines, macros and how to control data from being changed by using absolute cell referencing. Altering the display using formatting is a skill most students remember, but you might want to revisit using graphical displays.

Demonstrate a recap as required using the 'summer fair' scenario (Student's Book page 224) as the focus for the session. Start by working through the first two paragraphs under 'The summer fair'. Provide access to the Summer Fair file for students to use during the lesson.

Main lesson activities

Student task: Set **Activity 1** on page 224 of the Student's Book.

Student task: Move on to introducing functionality and formatting for currency (page 224). Ask students to follow the instructions given in the Student's Book.

Demonstration: Students may need some support with absolute cell referencing. It might be wise to set up a small spreadsheet to illustrate this. The spreadsheet could calculate total cost from adding tax to the cost price. Arrange that the rate of tax (say

10 per cent, which will be used as 10 per cent tax in the formula) is kept in a cell distant from the calculation and the calculation references this cell. When you normally copy the formula used in the calculation to other cells (relative cell referencing), the reference to the cell containing the tax amount is lost (the reference moves to the nest cell). You can demonstrate this, and then, as a comparison, use the $ sign to fix the reference to the tax cell. Remind students not to confuse the use of the $ sign with currency (Dollars).

Student task: Ask students to try **Activity 2** on page 225 of the Student's Book.

Give extra support by identifying problem areas and any students who are struggling. Often, a group of students will share the same problem, so it's worth spending some time identifying them rather than rushing round helping individuals separately. Once the group and the problem are defined, you can deal with the issue by demonstrating to the group and answering their questions.

Plenary suggestions

If this lesson is being delivered in two sessions, end this session with a plenary. This is an opportunity to share problems so that they can be dealt with in the next session. The formulae are given in this session, so they should not present a problem, but use questions to check that everyone fully understands the ideas of absolute cell referencing.

If you have time, ask the students to create their own scenario where absolute cell referencing would be needed. This could be a way of reinforcing the idea. This is also a suggestion for assessment – see below.

| **Assessment suggestions** | Suggest a scenario that will require some absolute cell referencing. Create a spreadsheet that demonstrates the operation of absolute cell referencing. |

5.8 cont. Data analysis: Spreadsheet modelling skills (double lesson)

Learning outcomes
- Incorporate formulae in spreadsheets.
- Use with understanding absolute and relative cell referencing.
- Alter the data in a spreadsheet so that a required outcome can be modelled.

Differentiated learning outcomes
- **All students must** be able to create a spreadsheet containing formulae and some appropriate formatting.
- **Most students should** be able to use absolute and relative cell referencing as required.
- **Some students could**, with understanding, use appropriate cell referencing and alter key data to meet a required outcome.

Resources
- **Student's Book: pages 224–228**

Starter suggestions

If this lesson is being delivered in two separate sessions, start with a recap on the skills covered in the previous session – in particular, absolute cell referencing.

Main lesson activities

Discussion: The importance and reasons behind testing should be discussed. It would be wise to work on both the current version and then saving it again with a different filename. The second copy can be used for the test data. Testing is best done with simple data so that the calculation is easy to check manually. Try changing some of the data; is it still giving mathematically accurate results?

If the model is returning mathematically accurate results, then students can use it in the modelling exercise below.

Student task: Set **Activities 3 and 4** on pages 226 and 227 of the Student's Book.

Discussion: Refer back to earlier work on absolute cell referencing and explain that naming cells works in exactly the same way. Point out the advantage of using this method if other workers are going to be responsible for maintaining the spreadsheet.

Student task: Set **Activity 5** on page 228 of the Student's Book.

This section is particularly important if you wish to keep copies of the students' work as printouts. It is also worth reminding students to use **Print Preview** and that altering the orientation to landscape is often useful. Finally, they need to print the formulae used on the spreadsheet in **Activity 6**; this often involves more adjustment of column widths.

Student task: Set **Activity 6** on page 236 of the Student's Book.

Give extra support by identifying problem areas and any students who are struggling. As suggested before, spend some time identifying students who share the same problem and helping them as a group, rather than helping individuals separately.

Give extra challenge by suggesting that students attach a macro to their spreadsheet – for example one that clears the values in cells C4 to C13 and D1, so that the model can be used again, maybe in another year. An extension could be that the macro first saves the spreadsheet and then clears the data.

Plenary suggestions

This is an opportunity to share problems so that they can be dealt with in the next session. Again, use questions to check that everyone fully understands the terms absolute cell referencing, testing and break-even.

Assessment suggestions	Here are two suggested assignments for the students to work on: 1 Show how you have tested a spreadsheet to make sure it functions to the intended specification. 2 Investigate the use of Goal Seek. Illustrate your findings with a suitable spreadsheet.

5.9 Developing your spreadsheet modelling skills

Learning outcomes
- Use and understand a number of functions including VLOOKUP, IF, SUM, MAX, MIN, AVERAGE, COUNT.
- Use HIDE as an aid to restricting the area printed.

Differentiated learning outcomes
- **All students must** be able to use several functions and restrict the area to be printed using HIDE.
- **Most students should** be able to use all of the functions covered in this session.
- **Some students could** show a full understanding of the functions met in this session, sufficient to be able to use them in a different context.

Resources
- **Student's Book: pages 229–232**
- **Files:**
  PPT5_9a.pptx
  MovieStars.csv

Starter suggestions

General point: This is a content heavy session and, depending on the experience of the class, you might need to run it as two sessions.

Using **PPT5_9a** show the class Shilpa's table of stars' names. Without referring to the Student's Book, ask students to guess the solution; then work out a couple of other (short!) names. They will realise that it would take ages to work out the whole solution; spreadsheet models can be used to solve puzzles and problems like this.

Main lesson activities

Discussion: Go through 'Problem solving with spreadsheet models' (Student's Book page 229).

Demonstration: Show students how to name a range of cells. This is an extension of naming a single cell from the previous session.

Demonstration: Follow the Student's Book to demonstrate the use of VLOOKUP to return the number 13 as a result of typing the function into cell D2. You could demonstrate the **Evaluate formula** facility in *Excel*, which enables you to 'step in' to each stage of execution of the function.

Discussion: When propagating the formula to cells I2 to N2, the point here is that with the name MIRZA, a good number of the cells will be empty since the name terminates in cell H1. This means that the function will try to match an empty cell within the named area A1 to A26; since there isn't an empty cell, the function returns the fault #N/A. Hence the use of the IF function.

Demonstration: Introduce the need for the IF function and demonstrate its action. The syntax for the IF function is given in the Student's Book on page 230; students need to understand fully how this operates. You could demonstrate how the function works outside this spreadsheet. Set up an IF function (for example, to return "Pass" if a mark is 60 or above and "Fail" if lower than 60). You could use:

=IF(cell reference>=60,"Pass","Fail")

where 'cell reference' is the cell containing the number to be tested.

Student task: Ask students to find and open the file **MovieStars**. Students need to keep a copy in their own directory; check that everyone has done so. Also check that students carry out the setting up of the named cells and the functions that start at 'Open **MovieStars.csv** ...' (Student's Book page 229).

Student task: Ask students to tackle **Activity 1** on page 230.

Discussion: Check for difficulties and successes. Be prepared to deal with student errors here – a common fault is missing brackets in functions. Point out why the completed function is called a *nested function*. The easiest way to copy the function between names (vertically) is **Copy** and **Paste** and then use **Fill** to copy horizontally.

Check that students remember how to use SUM and can work out the syntax for MIN, MAX and AVERAGE.

Demonstration: Show students how to add a footer to the spreadsheet. They could possibly add their name, file name, date, etc., to the footer.

Student task: Ask students to tackle **Activity 2** on page 231.

Student task: Implementing the COUNT function. The instructions for this are in the Student's Book (page 231), so there should not be too many problems.

Demonstrate: Show the students how to use **Hide** and **Unhide**.

Student task: Ask students to tackle **Activity 3** on page 232, following the printing operations instructions given in the Student's Book.

> Give extra support by demonstration; as mentioned above, it is a good idea to have a working version of the functions on a different spreadsheet (that is, using a different set of data).
>
> Give extra challenge by using the COUNTIF function to find the frequency of use of each letter of the alphabet.

Plenary suggestions

Review the lesson, particularly referring back to the use of the VLOOKUP function. Ask students what possible problems might arise if this function were copied to other cells. Ask students how this problem could be overcome.

Assessment suggestions	It can be useful to encourage students to record their progress in practical sessions by taking a sequence of screen shots. The shots can be cropped or reduced in size, then assembled as a series in a word processing document. The document helps greatly when assessing the merit of a piece of practical work.

5.10 Analysing spreadsheet data and preparing graphs (double lesson)

Learning outcomes

- **Understand and use the IF statement to display the appropriate outcome as a result of applying a logical test to data.**
- **Understand and use the nested IF statement to display a variety of outcomes as a result of applying logical tests to data.**
- **Use with understanding absolute cell referencing.**

Differentiated learning outcomes

- **All students must** be able to use with understanding the IF statement to display an outcome dependent on a logical test of the data.
- **Most students should** be able to use with understanding the nested IF statement to display different outcomes dependent on a logical test of the data.
- **Some students could** fully analyse the data using additional functions and charting tools.

Resources

- **Student's Book: pages 233–238**
- **Files:**
  MarkBook.csv
  MarkBookH.csv

Starter suggestions

General point: You may have introduced part of this lesson in the previous session when explaining (and demonstrating) the IF statement. In this session, the IF statement is used again, but here it appears as a series of nested IF statements.

Discussion: Demonstrate the **MarkBook** spreadsheet and ask the students how you could arrange for the spreadsheet to return either "Pass" or "Fail" as indicated on page 233 of the Student's Book. A supplementary question is: How does this expression work?

Student task: Ask students to find the **MarkBook** file, open it as a spreadsheet and save it in their own directory.

Try the conditional function as it appears on page 234 of the Student's Book.

Main lesson activities

Discussion and demonstration: How can the IF statement be modified to return "A star" in the Assignment Grade column where students have a mark of more than 90? Following the previous session, this should be easy for them to answer. Point out that the IF statement has three parts:

1. the "TEST"
2. what to do if the test is positive (in the above case it returns "A star")
3. what to do if the test is negative. In this case we don't need to fill this in because we only want to deal with the "A star" candidates.

What if we want to award "A" to candidates who obtain marks between 80 and 90? Here we can use the "unused" part of the IF statement. Ask the students what we should write in the third part of the IF statement. Expect some hesitant answers but accept ideas along the lines of using the "what to do if the test is negative" part as the location for a new test – in this case to test whether the score is between 80 and 90.

Demonstrate the new version of the function and draw the students' attention to the structure and how the name nested IF statement fits with the structure. Make sure that everyone understands this and then go on to show how the idea can be extended to the structure needed for an "A" grade. You should have something like this:

=IF(F2>=90,"A star",IF(F2>=80,"A",))

Now ask them to suggest how we can reference the data in columns A and B to give the relationship between grades and marks.

Students should be aware that absolute cell referencing is going to be needed. You could ask the supplementary question: Why use columns A and B? The answer would be that if the grade boundaries were to change, then changes made here (to columns A and B) will affect all of the students' grades.

Student task: Ask students to complete **Activity 1** on page 234 of the Student's Book.

Discussion: You will need to circulate and give assistance because there are many opportunities for errors here – notably because of syntax errors. In particular, watch out for missing commas, quotation marks and closing brackets.

> Give extra support by being prepared to demonstrate a previously debugged version of the spreadsheet and by giving guidance to small groups or individuals. Help any students who are really struggling by giving them a part-completed version of the function.

Plenary suggestions

Stop work and ask for comments on the work so far. Focus in particular on students who made mistakes but recognised and corrected their error.

You may wish to be sure that students understand and can use the nested IF statement. If so, you may decide to spend the whole of the first part of the lesson on this topic. You can resume with the use of VLOOKUP and HLOOKUP in the next session.

5.10 cont. Analysing spreadsheet data and preparing graphs (double lesson)

Learning outcomes
- Use with understanding VLOOKUP and HLOOKUP.
- Understand and use the functions COUNT, COUNTIF, MIN and MAX in the context of analysing data.
- Use formatting tools to alter the display of data.
- Use conditional formatting.
- Create, label and format the layout of a pie chart.

Differentiated learning outcomes
- **All students must** be able to understand and use the IF statement to display an outcome dependent on a logical test of the data.
- **Most students should** be able to understand and use the nested IF statement to display different outcomes dependent on a logical test of the data.
- **Some students could** fully analyse the data using additional functions and charting tools.

Resources
- **Student's Book: pages 233–238**

Starter suggestions

If this session is being delivered separately from the previous one, recap what was learnt there and explain that you will be building on that work to look at further ways of analysing spreadsheet data, including preparing graphs.

Main lesson activities

Introduce alternative ways of working using VLOOKUP and HLOOKUP (Student's Book page 234–235).

Student task: Implement the alternative ways of working using VLOOKUP, HLOOKUP and text wrapping, following the steps in the Student's Book, including **Activity 2**.

Demonstration: Moving to another sheet and renaming the sheet. Remind students of the use of the MAX, MIN and AVERAGE functions and how to name ranges of cells. Demonstrate these as necessary.

Discussion: How might the background colour of a cell be changed depending on its contents? Introduce the idea of an 'internal' IF statement where the user sets the conditions as a little piece of code that is attached to the cell and applied every time the cell contents change.

Student task: Work through the section 'Conditional formatting'.

Student task: Ask students to work through **Activity 3** on page 236. This should not pose too many problems as there are no new ideas to understand. Check that students have successfully completed the task.

Now move on to cover the COUNTIF function (Student's Book page 237). Some students may have encountered this already, but you need to demonstrate its use.

The syntax used in the Student's Book relies on the previous naming of ranges of cells, so check this has been done before starting.

Activity 4 focuses on the use of formatting to return percentage values. You may need to remind students how to format a cell or range of cells before they start.

Student task: Ask students to work through **Activity 4** (Student's Book page 237).

Discussion: Go over the material in 'Preparing graphs and charts' (Student's Book page 237). It is useful for students to be able to decide upon the most appropriate type of graph (chart) for data.

Demonstration: How to select data and how to create a pie-chart, ideal for representing percentages. Show the students how to format the display and how to change the title. They need to move the chart to a sheet in the workbook, so you may feel you need to revise how to insert a new sheet and rename the tab.

Student task: Ask students to work through **Activity 5** (Student's Book page 238).

Give extra support by monitoring students' progress; be prepared to offer part-completed functions as examples of what they should do. You can make the function address different cells so that the function cannot be directly used by the student, which means that the final spreadsheet keeps the integrity of being the student's work.

Give extra challenge by setting students a new assignment to use the logical functions AND, OR so that Mr Singh can identify those students who consistently produce work of a high (or low) standard.

Plenary suggestions

There is plenty to draw together after these sessions. It might be best to concentrate on the core of the work, which is the nested IF statement. A different example of its use may be a good way to complete the lesson, for example, (using the same spreadsheet) Mr Singh decides he wants to find out why students are not performing well in school. He may have data on the number of hours they spend on the computer playing games. The nested IF statement could, depending on the hours spent, give amusing results, such as =IF(cell reference>50," Time to turn off the console",IF(cell reference<1,"True?"...etc.

Assessment suggestions	Using the completed workbook as assessment material has its advantages, but the drawback is that students might submit work that is not completed under controlled conditions. It might be easier to use the Student's Book scenario as a practice piece and set a different context (such as a car salesroom or furniture shop) as the scenario for assessment covering similar activities.

5.11 Importing additional data and further work with graphs (double lesson)

Learning outcomes
- Import data in other formats into a spreadsheet.
- Select non-contiguous data.
- Use charting tools to format a graph.
- Make a simple sort data based on one field and a more complex sort based on two fields.

Differentiated learning outcomes
- **All students must** be able (with some help) to import data from a text file, select non-contiguous data, use chart-editing tools to format a graph and perform a sort based on one field.
- **Most students should** be able to work largely autonomously and alter formulae and functions as necessary. They should be able to perform a sort based on two fields. Their final print will largely be to specification.
- **Some students could** work autonomously and complete the tasks without error.

Resources
- **Student's Book: pages 239–243**
- **Files:**
  Names.txt
  NewGeographySummaries.txt

Starter suggestions

This is a full lesson, so we have suggested dividing it into two parts:

1 importing additional data and further work with charts and graphs (Student's Book, pages 239–240)
2 sorting and filtering data (Student's Book, page 242–243).

Discussion: Start with a discussion on file types. In particular, draw students' attention to the essential difference between text and a spreadsheet. In the spreadsheet, data is separate; the spreadsheet is made up of units of data in cells, each cell acting as its own text area, rather like a lot of pages of text all stuck together. To convert a text file to a file suitable for displaying in a spreadsheet, we need to tell the spreadsheet where each unit of data begins and ends.

Demonstration: Use the file **Names.txt** and show the students the text file in a word processor. Import the file into *Excel* and demonstrate the effect of using Tab (the default in the **Text Import Wizard**), semicolon, comma and comma + space. Comma works to separate most of the names, except for Mary and Robert; there is a space between the two names, so the space delimiter will work here. Depending on your spreadsheet software, you may need to experiment with the best import settings before demonstrating in lessons.

Five-minute activity: Students should practise importing the text file into their spreadsheet program. If there is time, they could alter the text file substituting different delimiters and then try importing the file into a spreadsheet.

Now students should import the file **NewGeographySummaries.txt** into their spreadsheet (with Comma and Tab used as the delimiters). Remind students to import to the active cell H1; if they haven't set this as the active cell, then they can select this in the last stage of the import. Check for general success.

Main lesson activities

Discussion: Before starting, mention that there are opportunities here for mistakes to be made, many of which are difficult to correct. It is best to save the working file regularly and possibly as a series of related file names. Briefly discuss the use of delimiters, as well as the changes needed to headings and the copy of the formula.

Student task: Ask students to complete **Activity 1** on page 239.

Activity 2 revises some of the work carried out in the previous session. In particular, some of the functions will need to be changed. It is probably best to run **Activity 2** immediately after **Activity 1** so that swifter students are not held back. Check for progress and give aid where necessary.

Discussion and student task: Preparing the bar charts following the instructions on page 240–241 of the Student's Book. This should be followed directly by the next task: completing **Activity 3** on page 241. The new skills here are the use of the CTRL key to enable the selection of non-adjacent cells, which are then used in the graph and the labelling of the graph axes.

Discussion: There are lots of alternative routes to creating and formatting charts in spreadsheet software. Discuss the alternative methods and shortcuts found by students.

Student task: Ask students to complete **Activity 4** (Student's Book, page 241), which again is familiar work.

5.11 cont. Importing additional data and further work with graphs *(double lesson)*

Learning outcomes
- Import data in other formats into a spreadsheet.
- Select non-contiguous data.
- Use charting tools to format a bar chart.
- Make a simple sort data based on one field and a more complex sort based on two fields.

Differentiated learning outcomes
- **All students must** be able (with some help) to import data from a text file, select non-contiguous data, use chart-editing tools to format a graph and perform a sort based on one field.
- **Most students should** be able to work largely autonomously and alter formulae and functions as necessary. They should be able to perform a sort based on two fields. Their final print will largely be to specification.
- **Some students could** work autonomously and complete the tasks without error.

Resources
- **Student's Book:** pages 239–243
- **Files:**
  NewGeographySummaries.txt

Starter suggestions

If this session is being delivered separately from the previous one, you may find it best to let students carry on from where they are, as students are likely to have reached different stages.

Main lesson activities

Demonstration: How to perform a sort. Point out that *Excel* assumes that the first row contains headings which, of course, are not to be included in the sort. Emphasise the need to select a range of cells and point out that if only one column were to be selected, then large errors to the data would result.

Student task: Ask students to follow the instructions on 'Sorting and filtering data' starting on page 242 of the Student's Book.

Discussion: Draw students' attention to the need for selecting all of the relevant data as mentioned on page 242 of the Student's Book (and by you before they started!).

Student task: Follow the instructions on page 242 of the Student's Book to make the formatting changes and the sort. Check the results for agreement.

Student task: Ask students to complete **Activity 5** on page 243 of the Student's Book. This is a print routine following another sort. Again, check that students are selecting all of the range.

Each student will need to print at the end of this full double session, so you may want to consider how to manage this if another class is expected in the room you are using. Make sure that the printouts can be identified with their author. For the students, the issue will be keeping their print to a single page.

Give extra support by circulating, identifying problems and offering advice. It is best to avoid making changes to a student's work. If possible, demonstrate what needs to be done, leaving the student to make the changes needed to solve the problem.

Give extra challenge by allowing students to work at their own pace. This is a session where some students will be able to progress rapidly, relying only on the Student's Book and with occasional reference to advice from you.

Plenary suggestions

Review progress. Ask the class about any difficulties encountered during the session, such as the need to select all of the appropriate data before making a sort. Mention that the use of the CTRL key for selecting non-contiguous ranges is a common facility in Microsoft (and other) office software.

Assessment suggestions	The printed text offers the opportunity to assess progress through the whole of the double lesson, but you will need to set a different exercise for assessment.
	The instructions given in the Student's Book are very precise – you will need to check that the skills and knowledge have been internalised by the students and that they can transfer their skills to other situations.
	To gather evidence of progress and applied skills, ask the students to make screen shots of important ephemeral evidence, such as the Sort dialogue box.

5.12 Data analysis: Referencing an external file and searching the data (double lesson)

Learning outcomes
- **Use the VLOOKUP function where the data to be referenced is held on a different spreadsheet file to that of the function.**
- **Use the SUMIF function.**
- **Use filtering tools to select data.**
- **Format the spreadsheet and print the sheet to specified criteria.**

Differentiated learning outcomes
- **All students must** be able to use, with help if necessary, the LOOKUP and SUMIF functions, select data using appropriate filters, and format and print their work to specified criteria.
- **Most students should** be able to use the LOOKUP and SUMIF functions autonomously and select data using appropriate filters.
- **Some students could** understand and use the LOOKUP and SUMIF functions and select data using appropriate filters, including the use of wildcards, in new situations.

Resources
- **Student's Book: pages 244–249**
- **Files:**
  IbanezTapas.csv
  SupplierCodes.csv

Starter suggestions

Discussion: Remind students of the use and syntax of the VLOOKUP function. Outline the first part of the lesson, where VLOOKUP is used to find data in one spreadsheet and copy related data to another. It is clearer for the students to follow your demonstration (below) if you open two instances of the spreadsheet and display these on the screen/whiteboard.

Demonstration/Student task: Demonstrate the two methods using VLOOKUP that appear on page 244. Then ask students to follow the instructions on that page.

Main lesson activities

Discussion: Check that the students have completed the use of VLOOKUP as above. Likely sources of error are problems with addressing another file (such as mistakes with the file name), forgetting to name the sheet correctly and adding an exclamation mark. Outline the tasks in **Activity 1** (page 245) and ask students for the name of the function to be used in no. 5. Discuss syntax used with COUNTIF.

Student task: Ask students to carry out **Activity 1**.

Demonstration: Outline the use and syntax associated with the function SUMIF and demonstrate the use of the function.

Student task: Check that students have correctly named range E16:E55. Check that everyone understands the syntax of SUMIF function. Then ask students to implement the use of SUMIF as shown on page 245 of the Student's Book.

Discussion: Check for success and discuss troubleshooting done by students. Refer to naming the range of cells and how these and cell A2 are used in the function. Discuss the formula needed in cell J16 and in K16.

Student task: Ask students to carry out **Activity 2** on page 245.

Demonstration: The first section of practical work on page 245 of the Student's Book requires students to use the SUMIF function again, twice. Show how the function locates the data from the range and the criterion set; once the locations of the data are identified, the "SUM" part of the function serves to total the values.

Discuss the syntax of the function again and check that everyone feels that they can make a start. Check progress and, if necessary, demonstrate the action of the function to small groups.

Discussion: Check on progress so far and respond to any difficulties met. Discuss the need to change the criterion used in the SUMIF function in cells D59:D61 (the selection criterion will need to change from "Pkt" to "Tin" and then to "Jar" and "Box" for the other cells). There is a similar problem for implementing this function in the range E59:E61, but it is probably best to leave the students to work this out for themselves.

Plenary suggestions

Review progress. Ask the class: If a function has to access data in a different spreadsheet, what changes need to be made to the function?

Accept the idea of 'a path… so that the function can find the data in the other spreadsheet', even if the students cannot remember the exact syntax.

Ask the class: How can we select specific data for summing together?

Here, the students should recognise that it is the SUMIF function that is needed, although others may remember the use of IF on its own and suggest a two-stage process of selection of data and subsequent addition.

5.12 cont.
Data analysis: Referencing an external file and searching the data (double lesson)

Learning outcomes
- **Use the VLOOKUP function where the data to be referenced is held on a different spreadsheet file to that of the function.**
- **Use the SUMIF function.**
- **Use filtering tools to select data.**
- **Format the spreadsheet and print the sheet to specified criteria.**

Differentiated learning outcomes
- **All students must** be able to use, with help if necessary, the VLOOKUP and SUMIF functions, select data using appropriate filters, and format and print their work to specified criteria.
- **Most students should** be able to use the VLOOKUP and SUMIF functions autonomously and select data using appropriate filters.
- **Some students could** understand and use the VLOOKUP and SUMIF functions and select data using appropriate filters, including the use of wildcards, in new situations.

Resources
- **Student's Book: pages 244–249**

Starter suggestions

Revise the use of SUMIF and the technique of referencing data that is found in a sheet other than the one where the main calculating is taking place. Have Senor Ibanez's spreadsheet, which has been built through **Activities 1** and **2** of the previous session, to hand in order to illustrate or inform this revision.

Main lesson activities

Student task: Ask students to complete **Activity 3** on page 246. The items beyond no. 3 are revision of previously visited techniques; depending on the class, you may need to refresh minds. Bear in mind the requirement to print in this task; this will be more of a problem if students are printing towards the end of the lesson.

Discussion: Introduce the need for filters to limit the data displayed on the screen. For instance, in the case of a large amount of data, such as in this example with many different producers and many different products, how can we find data we require? What if Mr Ibanez wants to find which products he obtains just from one supplier?

Demonstration: Show how to set up a filter following the dialogue on pages 247–248.

Student task: Ask students to carry out **Activity 4** on page 248.

Discussion: Review progress. Revise the use of wildcards.

Demonstration: Show the use of wildcards as documented on page 248.

Student task: Ask students to carry out **Activity 5** on page 248.

Give extra support by demonstrating to small groups or individuals.

Give extra challenge by suggesting that students investigate the effect of the wildcard "*". Can they make a rule for its use? Can they combine it with "?" to filter specific data?

Plenary suggestions

Review progress. Focus on the students' ability to search through the data in the sheet by asking more questions, such as: How many products come in a tin, and require 10 a day? Maybe ask them what the filter conditions would have to be in order to get the answer, that is, set "Pkt/Tin" to "Tin" and "Daily Quantity" to "Equals 10".

Assessment suggestions	The student activities could be used for assessment or you could devise a different scenario covering the same tasks as those in this session. Some ideas for the scenario could be: a car sales business, a firm specialising in bookings for different types of concerts, a mobile phone service provider.

6.1 File structure and terminology

Learning outcomes
- Describe what is meant by the terms 'file', 'record', 'field' and 'key field.'
- Describe different database structures, such as flat files and relational tables, and the use of primary keys and foreign keys.
- Describe the need and properties of relationships between database tables.

Differentiated learning outcomes
- **All students must** use the key words highlighted in the Student's Book in context and recognise that data can be organised and structured in different ways, suited to its use.
- **Most students should** confidently use the terminology in the session, especially the key words highlighted in the Student's Book, and state the differences between a flat file and a relational database.
- **Some students could** confidently use the terminology outlined in the session to describe concepts, not just being able to identify them; compare and contrast the differences between a flat file and a relational database.

Resources
- **Student's Book:** pages 254–256
- **Files:**
  PPT6_1a.pptx
  WS6_1a.docx
  WS6_1a_Answers.docx

Starter suggestions

Discussion: Ask students to outline the way that contact details in their mobile phone are organised. In discussion, emphasise the following:

- names will be in alphabetical order
- the same set of data items can be stored for each contact even though some items might be left empty
- each item must be referenced somehow by the software in the phone; for example, when you create a text message, the name of the contact is asked for and their number is automatically inserted
- a contact's details must also be recognised as a set of related data items.

Main lesson activities

Teacher advice: Emphasise the key terminology highlighted in the Student's Book for this session (file, record, field, record structure, flat file database, relational database, primary key, foreign key).

- In discussion about the breakdown of a file, use **PPT6_1a**.
- Student task: Set **Activity 1** on page 254. After students have created a field list for the music students, ask them to justify their choices.

Discussion: After reading through the text on key fields (Student's Book, page 255), expand the list beginning with ISBN and telephone numbers, with other unique data items that might be used as key fields. Suggestions might include Examination Candidate Number or seat number in an aircraft seating plan. Set **Activities 2** and **3** on page 255.

Teacher advice: After working through the car salesman example (section on 'Flat file or relational database') and before starting **Activity 4** (Student's Book, page 256), ensure that the ideas presented so far are fully understood. Perhaps work though **Activity 4** as a class activity.

> Give extra support by drawing on the mobile phone contact structure. Draw this out and see if links could be made from the contact structure to a structure for their mobile's text inbox.
>
> Give extra challenge by:
> - posing questions such as: within the orchestra database what would happen to the record table if a student were to wish to learn more than one instrument?

Plenary suggestions

Supply some examples of databases. Ask students for a list of necessary fields and then identify the primary key field.

Student closing activity: Ask students to work in pairs to complete **WS6_1a**. (Answers provided: see **WS6_1a_Answers**.)

6.2 Data-handling situations

Learning outcomes
- Describe different database structures, such as flat files and relational tables, including the use of relationships, primary keys and foreign keys.
- Describe the use of database structures within the finance industry the library system.

Differentiated learning outcomes
- **All students must** use the key terminology in context and recognise that data can be organised and structured in different ways, suited to its use.
- **Most students should** be able to describe the use of a flat file or a relational database in context and suggest record structures and relationships.
- **Some students could** confidently evaluate the use of flat files and relational databases in given situations, highlighting and justifying structures and relationships.

Resources
- **Student's Book:** pages 257–259
- **Files:**
  WS6_2a.docx

Starter suggestions

Ask students to think about the receipts that they are given at any retail outlet. The product details are printed, not just the price. A barcode will identify the product, so where do the other details come from? As students are making suggestions, ask one student to act as a scribe and draw what is developing on the board. Have students developed a flat file or a relational database? Why?

Main lesson activities

Student task: Ask students if they have seen someone at home using the internet for purchasing food, CDs, DVDs or books. What kinds of details are needed for that activity to be successful? Can students make a justified (tentative) record structure using those details?

Hold a Q&A session: Mention that relationships are not just 'one-to-one' (an immediate assumption when using multiple tables is simply to split up a large single table), but that 'one-to-many' relationships are usual. For example, one supplier will provide many products. (SupplierID may be the primary key to a 'Supplier' table and may appear in many product records in the ProductList table, in order to stop duplication of data.)

Discussion: Work through the section on 'Insurance' (Student's Book, page 257) and talk through the process that is taking place with the database that Noura's father is using. Explain why there might be external links (such as to ServicePlus). Ask students to suggest what external links a supermarket might have in a product's record.

Teacher directed discussion: Move on to the section detailing the data processing that takes place in a library. Discuss the need for the three tables in order to record the borrowing of a book: the fields required and the purpose of the table **Borrowed**. It is important that students understand the fact that the details in **Borrowed** are only held temporarily (for the duration of the book being borrowed).

Discuss the relationships and emphasise the 'one-to-many' nature of them. It might be necessary to recap why **Borrowed-Member** is not 'one-to-one.'

Task: students complete **Activities 3** and **4** (Student's Book, page 258).

> Give extra support by concentrating on the library system. Bring students together and then convert the description of the borrowing procedure into a flowchart. Identify places in the routine where problems could occur.
>
> Give extra challenge by considering how the music students' database considered in Session 6.1 could be linked to other school systems, such as the student registration system.

Plenary suggestions

Ask students to present to the class their answers to **Activities 3** and **4**. Try and ensure that the 'Return' routine is class-approved and workable.

Assessment suggestions	The answers to Activity 3 (and Activity 4, if completed) will provide good evidence of understanding. Watch especially for how justifications are suggested and received.

6.3 Data types

Learning outcomes
- Identify different data types: logical/Boolean, alphanumeric/text, numeric (real and integer) and date.
- Select appropriate data types for a given set of data.

Differentiated learning outcomes
- **All students must** be able to identify alphanumeric/text, numeric and date as appropriate data types for a field.
- **Most students should** be able to choose and describe appropriate data types for fields.
- **Some students could** choose the appropriate data types for fields, justifying selection.

Resources
- **Student's Book:** pages 260–261

Starter suggestions

Ask students to make a list of 10 attributes about themselves, such as age, height, hair colour and gender. For each of these, ask them to describe the type of data that it is: a word, a number (whole or fractional), a date and so on. Now ask them to consider the processing that each of these might need, such as:

name – sorting alphabetically

height – converting from imperial to metric measure

date of birth – calculating age.

Main lesson activities

Student task: Assign data types from the table on page 260 of the Student's Book to the 10 attributes listed in the starter. Can they have an item of data for each data type? Currency – pocket money?

Teacher advice: With **Activity 1** on page 261, spend a fair amount of time exploring the possible data types. This is important because students need to understand the data types fully (especially any different settings possible) in order to make the best choices when creating a record structure in a new database. Note that it is sometimes very difficult to make changes after data has been added into a database.

Also consider that data items such as car registrations are text (containing letters as well as spaces and numbers); a telephone number may be a text field (if it has brackets to indicate a country or area code; hyphens). If a telephone number is set as a number data type, any leading '0' will be removed.

Teacher advice: **Activity 2** leads to a complete and usable record structure. Check for meaningful names. Allow students time to justify their data type choices. Ask them to give examples alongside each field as in the table in the text.

Give extra support by going back to the initial list of personal attributes they made in the Starter. Consider data types again now that students have been properly introduced to them. Have a look at the record structure for the cars and books from Sessions 6.1 and 6.2.

> Give extra challenge by asking students to consider how dates might or should be stored so that when they are sorted they appear in chronological order.
>
> What items of their personal attribute would be real numbers or integers?
>
> If they have chosen height or weight to be measured by imperial units and stored as real numbers, can arithmetic be done on them, such as calculating body mass index (BMI)? Would there need to be some conversion, for example, from imperial to metric or decimalising an imperial measurement (10st 7lbs → 10.5 stone).
>
> Does any of the data have a particular character or numerical size (height will be less than 2 m, what characters could be used to represent gender?)

Plenary suggestions

Ask students to look at the three tables that were given as the answer to **Activity 2** in Session 6.2 and to create a full record structure for each of them.

Assessment suggestions	Successful completion of Activity 2 and confident verbal answers to the plenary session will indicate that the student has a good understanding of the ideas presented in this session.
	Ask students to produce, unaided, a record structure for their 'personal details'. Some students will be able to add information about a field's size and data range.

6.4 Searching and querying

> **Learning outcomes**
> - Sort data into ascending or descending order using different field types.
> - Search or select subsets of data using multiple criteria, different field types, a variety of operators and wildcards.

Differentiated learning outcomes	Resources
All students must be able to sort and search a data set with single, given criterion. **Most students should** confidently sort and search a data set with single criterion and perform logical condition and wildcard searches with given criteria. **Some students could** confidently use operators, logical conditions and wildcards to sort and search a data set.	**Student's Book:** page 262–263 **Files:** WS6_4a.docx WS6_4a_Answers.docx

Starter suggestions

Students should consider their music or book collections. Could they sort their CDs or books into alphabetical order by performer or author? What happens if they have more than one CD or book by the same performer or author? Introduce the idea of sorting within a particular grouping. How would they find CDs or books with a particular word in the title?

Main lesson activities

Student task: Ask students to do **Activity 1** on page 262 to ensure that they can perform the sort and get the same results. Then ask students to practise putting the eight car records on page 262 of the Student's Book into different orders. Remember that to sort always returns the full data set, whereas to search returns only the records that fulfil that search criteria.

Student task: For an extra activity, create a set of records from the personal details of students from the Starter activity of Session 6.3. Ask students to sort the data into various orders. Start a competition for individuals or teams to complete, sort and correctly report the new order.

Teacher advice: **Activity 3** builds directly on **Activity 2**, so it is vital that students fully understand the answers to **Activity 2** before moving on.

Q&A: Before setting **Activity 3**, give students practice with some similar searches, such as Opel cars with 100 000 miles or more, or registrations beginning with 'A' that cost more than $1500.

As another activity following on from the section on wildcards, use the data created about the class and ask questions of that: students with surnames Like "S*", students with DoB Like "??/06/??" and so on.

Student task: Ask students to complete **WS6_4a** – either in class working in pairs or (if pressed for time) as a homework task. Answers are provided on the Teacher Guide digital download – see **WS6_4a_Answers**.

Give extra support by photocopying the list of cars in the text, chopping the table up into rows and then rearranging the 'records' on the desk. Sort into different orders of fields within fields.

Give extra challenge by asking students to produce searches on the class data using logical conditions and wildcards. Once this has been done, ask them to produce the query language statements that will achieve these searches.

Plenary suggestions

- Play a game of search in reverse. Ask students to come up with the query statement that will produce the answer that you give. For example, using the car records, you might have an answer of 'Registrations are: JFA 857 and CAA 876'. The search would have to be: Like "??A*"

Assessment suggestions	All students should be able to apply a search, so make sure that everyone can apply at least one criterion. Most should have managed to complete all of Activity 2 successfully. Some students should be able to complete Activity 3 successfully.

6.5 Creating a database

Learning outcomes
- Define an appropriate database record structure.
- Enter and amend data in a database.

Differentiated learning outcomes
- **All students must** be able to import data with field names from another source.
- **Most students should** be able to import data with field names from another source; add, amend and delete records.
- **Some students could** import data from another source (with or without field names), and add, amend and delete records.

Resources
- **Student's Book:** page 266–271

Starter suggestions

Revise the terminology of Session 6.1 (record, field, key field, table) by referring to any of the structures that were created or explored in Sessions 6.1 to 6.4.

Main lesson activities

Discussion: The concept of an empty database being saved can sometimes seem strange when our usual experience is to save at a particular stage of our working. Make the point that a space is being reserved where data that we upload can be placed and that as we type in new data it is automatically saved – we do not have to keep on resaving at intervals.

Sometimes organisations have naming conventions for objects in the database as well. For example, a table name has the prefix 'tbl,' a report name has the prefix 'rpt'. This would produce **tblStudentDetail**. If students are familiar with this or you have a convention, then follow that, but keep the object names used in the text after the prefix.

Teacher advice: It is not necessary to enter the comments into the record structure – they are there for reference only. Make sure that students have the correct settings for the data types, and that 'AdNo' is set as the key field.

Student check: Before starting the section 'Editing: adding, changing, deleting' on page 268, ask students to check that the data types match what is needed by looking at the design view of the record. This is especially true when importing data that does have field names included.

Teacher advice: All students should be able to reach the end of **Activity 3** successfully. It is vital that from this point onwards all students have a full and accurate version of **MusicStudents**, so it is advisable to have a version prepared that can, if necessary, be shared.

Give extra support by spending longer over, or repeating, the session up to and including Activity 1, ensuring that all students can import data from an external source.

Give extra challenge by asking students to do the following:

- Create a spreadsheet of the car records in Session 6.4 with the field names in row 1. Save the sheet as a .csv (text) file. Create an empty database and import the .csv file.
- Add a character size to the text fields in the record structure. Students will have to study the data carefully to work this out.

Plenary suggestions

Break up the task of creating a new database and importing data into distinct steps and have each step written on separate pieces of paper/card. Ask students to rearrange them into the correct order.

Assessment suggestions	All students should complete the session up to and including Activity 3 confidently. How well students manage Activity 4 will allow you to make a judgement on the level of understanding they have achieved.

6.6 Creating relationships

Learning outcomes
- Import data from .csv files into new tables.
- Create relationships between tables in a relational database.
- Enter and amend data in specific tables.

Differentiated learning outcomes
- **All students must** be able to import data with field names from another source; be able to create a relationship between specific fields.
- **Most students should** be able to import data with field names from another source; add, amend and delete records; decide what relationships can be created between tables.
- **Some students could** import data from another source (with or without field names), and add, amend and delete records; be able to discuss the type of relationships that can be created between tables.

Resources
- **Student's Book:** pages 272–275

Starter suggestions

Revise the terminology and ideas of relationships that were discussed in Session 6.1 (record, field, key field, primary key, foreign key, table, relationship) by referring to any of the structures that were created or explored in Sessions 6.1 to 6.4.

Main lesson activities

Discussion: Revise the steps needed to locate and import the two .csv files (look for text files in the folder) into two new tables. Ensure that the fields have the correct data types assigned, and that the correct fields are set to be primary keys.

Student task: bring the two .csv files into the database ensuring that the data types, key fields and Case set to Yes/No are correct. Carry out the three additions to the files. Make sure that the number of records in each table is correct.

Teacher advice: All students should be able to reach the end of **Activity 1** successfully. It is vital that from this point onwards all students have a full and accurate version of **MusicStudents**, so it is advisable to have a version prepared that can, if necessary, be shared.

Discussion: It is important that students understand what they are going to achieve in the next few steps, so spend a few minutes exploring the relationship 'map' on page 273. Revise the relationship types emphasising that 99 per cent of all relationships they will build are 'one-to-many.' Be sure to clear up any confusion as to why the relationship between **Instrument** and **StudentDetail** is 'one-to-many' and not 'one-to-one.'

Teacher demonstration: Walk through finding steps of finding the area where relationships can be created, the adding of tables to the area and then actually creating relationships between the correct fields in the tables. Ensure that students understand and can interpret the information in this relationship diagram.

Student task: Work through these steps to create the two relationships between the three tables. Ensure that the database is saved in its new state. Have a look at the way the records are now displayed in the tables, exploring the connections that have been made between the records as shown on page 274.

Carry out **Activity 2**. Check the corrections have had the desired effect.

Discussion: save students in small groups working out an answer to **Activity 3**. Gather the justifications together on the board and summarise them.

> Give extra support by spending longer over, or repeating, the session up to and including Activity 2, ensuring that all students can import data from an external source and create relationships between tables.
>
> Give extra challenge by asking students to create and link the table in Activity 4

Plenary suggestions

Have a discussion about the advantages and disadvantages of building relationships between tables to create a relational database instead of using a single flat file. Could all the data that is now in the database have been put into the original **StudentDetail** table? Draw all this together on the board.

Assessment suggestions	All students should complete the session up to and including Activity 2 confidently.
	How well students manage Activities 3 and 4 will allow you to make a judgment on the level of understanding they have achieved.

6.7 Creating a data entry form

> **Learning outcomes**
> - Create a data entry form that matches the requirements of the user.
> - Create a data entry form that has appropriate data capture fields.

Differentiated learning outcomes

- **All students must** be able to create a data entry form which has all the fields of the record to be added.
- **Most students should** be able to create a data entry form that makes good use of the space on the form and highlights key fields, or offers information for the user.
- **Some students could** create a form that meets the user's requirements, including good use of white space, offers validation, and information to ensure 100 per cent accuracy.

Resources

- **Student's Book:** page 276–279

Starter suggestions

Display on the board some examples of data entry forms on websites. These could include forms that are submitted to ask for additional information (for example, to request a school prospectus), or are to submit information to a comparison website (like Noura's father in Session 6.2). Discuss what makes the forms attractive and effective. Explain the importance of clarity and simplicity. Are there any checks to ensure data entry is as accurate as possible?

Main lesson activities

Discussion: The task is to create a data entry form for the table **StudentDetail**. On the board bring together student ideas about:

- any checks on accuracy that could be made
- information that might usefully be placed on the form to help the person who is supplying the data
- how the field should be organised or collected together into areas.

This will help to focus attention on how the design of the form should take shape. Leave this information visible on the board.

Task: walk students through locating the button that gets them into the form wizard and connecting the form to the table **StudentDetail**.

Once the basic form is displayed ask students to save it.

Discussion: you have already encountered validation in Session 6.5 (as well as when working with spreadsheets in Session 5.8). What fields here might be suitable for similar validation? Orchestra, Own are obvious choices for Yes/No type validation checks (Session 6.5). There is only a limited number of Teachers – they could be in a list like the Tutor Groups are (Session 6.1). This is a good start.

Teacher task: explain the idea of a drop-down list, what it looks like and how this can be connected to the Teacher field.

Teacher demonstration: go through the steps of creating a drop-down list on the form for the Teacher field. Ensure that students have followed this fully.

Student task: copy what the teacher has done. Follow the steps in the Student's Book (page 269). Once students have the Teacher drop down list working properly, they can try **Activity 1**.

Teacher task: this has been a big step that may have caused a few problems. Bring the class back to focus on the next task – putting onto the form radio/option buttons.

Explain what radio buttons look like, how they work. The key here is an explanation of how two buttons, one for 'Yes' and one for 'No' will be grouped together to answer the question, "Does the student have a school instrument?"

Teacher demonstration: go through the steps of creating the option group with the two buttons in it. Again, ensure that students follow this fully.

Student task: copy what the teacher has done. Follow the steps in the Student's Book (page 271).

Discussion: talk about the layout of the form, its attractiveness, its function, its ease of use, and so on. Refer back to the ideas collected together on the board from the earlier discussion.

Student task: complete **Activities 2** and **3**.

The form should now be fully functional.

Student task: complete **Activity 4**.

Give extra support by walking students through the creation of the two drop down lists on their own form. Do not be too concerned about them having radio buttons as well. They have successfully added a means of validating and ensuring that these data items are 100 per cent accurate.

Give extra challenge by asking students to create an option group with radio buttons asking the question, "Is the student an orchestra member?"

Plenary suggestions

Have a discussion about the advantages and disadvantages creating a data entry form to enter new details rather than have users adding directly into the table.

Assessment suggestions	If students can successfully complete these activities, they will have learnt a terrific database building skill – especially if they can do so confidently and with minimal prompting. All students should be able to at least create the form
	How well students were able to contribute to the Plenary session will also give you a good idea of their level of understanding of the need for data entry forms.

6.8 Filters, queries and creating reports

Learning outcomes
- Sort data.
- Search data to select subsets of data.

Differentiated learning outcomes
- **All students must** be able to create a report from a sorted data set.
- **Most students should** be able to attach a query to a report.
- **Some students could** create a query with one criterion and attach it to a report.

Resources
- **Student's Book:** page 280–283
- **Files:**
  WS6_8a

Starter suggestions

This session puts into a practical context the question framing in the early part of Session 6.4. Revise the concepts of sorting and searching from Session 6.4. Look at the questions in Session 6.5, **Activity 3** and consider how we might ask the database to return the answers we expect. What questions should be asked?

Main lesson activities

Teacher advice/points to note:

- Spend some time making sure that students understand the difference between sort and search, as well as the phrase within.
- Make sure you know where the buttons for **Filter**, **Advanced filter** and **Query** can be found.
- Make sure students know how to move between the design view and the data view, or between the filter settings and the data view, so that they can make adjustments and corrections.
- Be sure of the difference that your software makes between filters and queries, particularly with regard to saving them – and where that option can be found.
- Stick with creating reports through the report wizard. Even after looking at the make-up of a report in the next session, it is a highly complex activity to produce a report in its design view.
- Layout design or style is not, at this stage, important. The important aspects are selecting the correct data and then displaying it in the report so that all data fields and headings are visible. The only layout aspect to practise is that of orientation: portrait or landscape (see **Tip**, page 282).

Demonstration: Give a demonstration (if necessary) of Task 1 of **WS6_8a**, and then ask students to complete Task 2 themselves. Alternatively, if students are sufficiently competent, they can follow the explanation on the Worksheet and complete Tasks 1 and 2 themselves.

Give extra support by producing several reports from filtering (sorting) the database – reports created from the whole dataset that has been reordered in some way.

Give extra challenge by asking students to use the small car database that they might have created through the last session. Ask them to add several more cars to the database (either using their imagination, looking at a car sales website, or using data previously prepared by you). Then ask them to create queries to select a subset, and attach them to reports.

Plenary suggestions

Q&A: End the session with the Task 3 of **WS6_8a**.

Set sort and search questions for students to answer using the **MusicStudents** database. For example:

If we sort in descending date of birth, who is at the top of the list? (David Mountain)

How many students does Mr J Hill teach? (5)

Assessment suggestions	You will need to monitor carefully how successfully students manage to complete all three Activity tasks in this session, ensuring they can produce a report. The plenary session will give an excellent indication of how well students understand how to question the database.

6.9 Reports and run-time calculations

Learning outcomes
- Use arithmetic operations or numeric functions to perform calculations.
- Search data to select subsets of data.
- Produce reports to display fields and other data for a user.

Differentiated learning outcomes
- **All students must** be able to create a report from a sorted data set.
- **Most students should** be able to attach a query to a report and create a run-time field.
- **Some students could** create a query with one criterion and attach it to a report; create and format a run-time field.

Resources
- **Student's Book:** pages 284–288

Starter suggestions

Look closely at the **StudentsByAdNo** report produced in Session 6.8. Can students identify elements such as headers and footers? Pose the question of having totals running from one page to another. Would those totals appear in the footer already identified? Why/why not? What else, if anything, might be needed? Try and steer students towards seeing the need for a report footer as well as a page footer.

Main lesson activities

Teacher advice/points to note:
- Spend time making sure that students fully understand the difference between page footers and report footers (Student's Book page 284). The crucial difference is how many times each footer will be displayed – the same applies to page headers and report headers as well.
- Take time to allow students to see the connection between a report's layout and its print view. The confusing aspect may be the repeated **Detail** element since it appears only once in the design view.
- You may not see any space for the report footer in which to add new fields and may have to drag down some space. Make sure students know that these actions are sometimes needed. The wizard simplifies report building and expects only that a field might need editing, or formatting or repositioning in the existing space.
- Yes/No should be displayed in a Boolean field, not 0/1 (or a tick box). If Yes/No is not seen in the print view, it means that the setting in the record structure needs adjusting. Yes/No is the data type, but it has the wrong setting. In the record structure, the setting should be: text box.
- When typing in query criteria, values like "Yes" or "Miss N Boyd", must be exactly as will be found in the field concerned. In these cases, neither "Y" nor "Miss Boyd" would return any records.
- Make sure students are clear about the way in which text boxes and labels work. Typing a calculation into a label will not work as intended – it will simply display the calculation as a label.
- The text box for the average cost per lesson calculation will need formatting. Your software will not default to a currency format with two decimal places –

you have to make the change. Make the changes using the **Properties** for the text box.

Give extra support by adding: =Count([AdNo]) into both the page and report footers of the StudentsByAdNo report to illustrate the different effect of the two footers. This will give a count of the students on each page in the footer of each page, and a total count of all students after the last student on the last page.

Give extra challenge by asking students to create a report from their car database that lists a certain car make (or colour or price range) and uses a count function (how many cars) and a sum function (value of the cars in the list).

Plenary suggestions

Hold a Q&A session:
- What function would I need to …?
- Which field(s) would the function need?
- What type of control would be needed to add in a …?
- Where in the report layout would I put a …?

Assessment suggestions	This session covers some high-level tasks, which may be hard for students to complete successfully. An awareness of what a run-time function is should be expected of most students, so the question/answer session will highlight that. Students with a strong command of the topic should be able to give verbal answers to the plenary questions with confidence.

6.10 Complex queries and wildcards

Learning outcomes
- Search data to select subsets of data.
- Base searches on using multiple criteria, arithmetic or logical conditions, wildcards.
- Produce reports to display fields and other data for a user.

Differentiated learning outcomes
- **All students must** be able to apply logical operators.
- **Most students should** apply logical and arithmetical conditions and wildcard searches to a data set.
- **Some students could** sort and search for multiple criteria; develop and apply a range of operators, logical conditions and wildcards to searches.

Resources
- **Student's Book:** page 289–292
- **Files:**
  WS6_10a.docx
  WS6_10b.docx
  WS6_10b_Answers.docx

Starter suggestions

Revise the concepts introduced in Session 6.4 (operators, logical conditions and wildcards).

Main lesson activities

Discussion: Spend some time making sure that the difference between the results for the AND, and the OR, 'no violin' searches is fully understood. When multiple criteria are applied, it helps to have an idea about how many records are likely to be returned. For students who do not own a violin, we would never expect 33 as an answer since there are only eight violin students.

Demonstration: Try another mix of AND with OR, such as: male drummers or female pianists (returns five students). As stated above, it is important for students to understand – and to be able to gauge the difference between – these searches, either separately or as a mix.

Student task: Students should now try **Activities 1** and **2**, and other queries using AND/OR with NOT. For example: all students (except those of Mr A Atkinson) who do not have their own instruments (19 students).

Demonstration followed by student task: Demonstrate some more wildcard queries. For example: students born in the calendar year 1996 (14 students), students with an AdNo year 1996 (18), students who have male teachers (21). Students should then try **Activity 3**.

Discussion: Revise adding queries to reports, adding run-time calculations, and using report footers.

Student task: Give students **WS6_10a** and ask them to read the explanation, investigate syntax and then complete the task.

Student task: Students to work in pairs to complete **WS6_10b** (or use as a homework task). Answers are provided – see **WS6_10b_Answers**.

Give extra support by practising single AND and OR searches, such as:
male violinists (4), pianists or drummers (9), students of Mrs M Potts who have their own instruments (5).

Give extra challenge by asking students to apply some of these ideas to their car database, for example: red Fords; built after 2003 or a mileage less than 50 000; has a registration with the letter 'A' somewhere within it.

Plenary suggestions

This is a session that really needs to have a game played at the end. There are two good options:

- ask students to produce queries to return the number of musicians that fulfil particular multiple criteria
- give a specific answer and ask students to work out what search criteria will produce that result. Students will sometimes produce different criteria that generate the same answers so also have a look at seeing who can produce elaborate search criteria.

Assessment suggestions	You will need to monitor carefully how successfully students manage to complete session activities. The 'Main lesson activities' suggested above will give a good idea of the level of understanding.
	As with Session 6.9, this session contains a difficult set of concepts for some students. All students need to understand the application of arithmetic operators, and the most confident should be able to deal with AND, OR and NOT criteria. Judicious monitoring and questioning of students during the plenary session is needed.

6.11 Producing labels and business cards

Learning outcomes
- Sort data.
- Search data to select subsets of data.
- Produce a label set from a subset of data.

Differentiated learning outcomes
- **All students must** be able to search for a correct subset of data.
- **Most students should** be able to create labels for a whole data set.
- **Some students could** create labels for a specific subset of data.

Resources
- **Student's Book:** pages 293–297
- **Files:**
  WS6_11a
  WS6_11a_Answers

Starter suggestions

Ask students to imagine that they work for a mail order company. They need to produce address labels for all the parcels that they are posting out. Would they hand-write them? Would they print out a list and then cut them up and glue them onto the parcels? No. Have some labels from packages available. From what they know about producing reports, do students have any idea how this might be achieved? You are looking to elicit something on the lines of, 'instead of details in a row, details are in a box' – something to indicate that students are identifying that the fields of one record are displayed in a new format.

Main lesson activities

Teacher advice: Judge the level of your group prior to this lesson. Some students will need a lot of guidance, but others will be able to follow the instructions themselves from the Student's Book. It might be a good idea to run through creating labels from the entire dataset as a walk-through demonstration.

Teacher-led student task: Take some time to look at the label options that your software offers you. Explore the sizes, the layout across the page, and any custom size option. Ask students to think of three or four useful labels that could be produced with these different sizes. Consider how much information can be fitted into the space that a label offers.

Demonstration: Revise the control panel and the placing of a label.

Discussion: What other reasons might there be for producing a series of labels for the whole set of music students? Students may be given a school folder to keep music in, needing an identity label. All music lessons take place at lunchtime so all students need an early lunch pass – a label attached to a small postcard.

Demonstration followed by student task: It will be important to discover how to make the rectangle transparent. Allow students to play around with the rectangle (and line) controls – not just transparency, but colour is needed in **Activity 3** (Student's Book page 296). When you place a rectangle onto the label, it is a layer on top of the information already there. Alternatively, you could look at the layer order and push the rectangle to the back.

Student activity: Complete **WS6_11**. Note to teacher: you will need to decide what tasks can be completed by which students. Answers are provided – see **WS6_11a_Answers**.

Give extra support by setting up labels printed in a different order: alphabetic FamilyName order; AdNo order… Introduce labels, images, rectangle and line controls.

Give extra challenge by asking students to:

- create labels that could be used on the instrument cases of the students taught by Mr A Atkinson who do not have their own instruments (10 students). Place suitable images and controls onto these labels.
- use word processing skills to produce a "How do I create labels?" user guide.

Plenary suggestions

In this session students need to be able to identify when the printed-label option could be used appropriately and recognise the steps needed to create the labels. This means that questions need to be asked to reinforce these necessary steps. For example:

- Can you select students after the labelling process has begun? (No)
- Can footers and headers be placed on a label? (Yes)
- Could one label be a whole page? (Yes)

Assessment suggestions	Successfully producing the required labels from the whole dataset (from table and not query) is a basic requirement. Most students should be able to produce a label set from the whole dataset with suitably arranged fields plus labels. Some students will be able to produce a query for any subset and then produce the required labels.

6.12 Summarising data for use in other software

Learning outcomes
- Search data to select subsets of data.
- Produce reports to display fields and other data for a user.
- Export data that can be used in other formats.

Differentiated learning outcomes
- **All students must** be able to search for a correct subset of data.
- **Most students should** be able to add a summary field to a query.
- **Some students could** create a query with multiple summary fields; export data in a format that can be used in other applications software.

Resources
- **Student's Book:** pages 298–301
- **Files:**
  WS6_12
  WS6_12_Answers
- **Further resources:**
  Brochures and reports containing text with charts and graphs

Starter suggestions

Have some brochures or reports available that contain charts or graphs within the body of text. Work backwards – ask students how and where these charts and graphs could be produced? what data they would require? what summarising questions (query) might be needed to make them? This leads to the question of how the data is exported from the database after the summaries have been made. Show how data flows between these stages and applications. This can be a good interactive opener with plenty of scope for student participation. It is well worth spending time here as this is a tricky session.

Main lesson activities

Teacher advice: It would be wise to plan on this session being a very interactive lesson, with more demonstrations than the previous sessions.

Make sure that you know where the **Totals** element of the query can be made visible. Explain to students the effect of the options available through the **Totals** element.

Demonstration: Demonstrate on screen with a walk-through of the task on page 298 of the Student's Book (making sure that everyone arrives at the table shown in the text and that the correct information is exported).

Discussion: Revise what happened in Session 6.5 with the .csv files. A .csv file is a kind of text file particularly suited to storing information in columns. This is a good file type for exporting data that will be loaded into a spreadsheet where charts and graphs can easily be created. A drawback of a .csv file is that no formatting can be stored in the text (bold, italic). Nor, for example, could tables have fill colours in cells. In order to save with some formatting, then an .rtf file needs to be used.

Teacher advice: Once a single summarising criterion has been achieved, avoid moving onto adding another until the process is fully understood.

Student task: Judge how much direction will be needed to enable students to complete Task 1 of **WS6_12** by themselves. This could be used as an additional teacher demonstration.

Student task: Students who are able to complete Task 1 of the worksheet could then move on to Task 2.

Give extra support by summarising with a single criterion over as many fields as you can think will show meaningful results: sum of lessons for males; sum of money paid each week by tutor group, and so on.

Give extra challenge by asking students to create some summarising tables from the cars database, for example, value of cars by make.

Plenary suggestions

Show some pre-prepared charts and graphs from **MusicStudents**. Ask what data would need to have been summarised in order to create that chart or graph. Then move on to working out what the query would be and what summarising total is being used.

Student task: Ask students to work through Task 3 from **WS6_12**. Answers are provided – see **WS6_12_Answers**.

Assessment suggestions	Tasks 1 and 2 of Worksheet 6.12 provide good indicators of success: <ul><li>All students should be able to produce the query for Activity 1 and with support, identify all the true statements in Activity 2, although not necessarily give the alternatives to the false or impossible ones.</li><li>Most students should be able to set up the query for Activity 1 and answer most of Activity 2 successfully.</li><li>Students with a strong command of the topic should be able to answer both these questions confidently and fully.</li></ul>

7.1 The systems life cycle: Analysis

Learning outcomes
- Recall and understand methods used to investigate a system.
- Select a method of collecting information relevant to the investigation.
- Identify the requirements of a new system: Information, hardware and software.

Differentiated learning outcomes
- **All students must** be able recall and understand methods used to investigate a system; understand the need for recording and analysing the existing system.
- **Most students should** be able to select a method of collecting information relevant to the investigation; be able to describe the recording and analysing of information connected to the existing system; be able to identify elements of a new system.
- **Some students could** provide convincing reasons for their choice in selecting a suitable method for collecting information; are able to describe the recording and analysing of information connected to the current system; be able to identify and justify a new system specification.

Resources
- **Student's Book:** pages 306–308
- **Files:**
  PPT7_1a.pptx
  WS7_1a.docx
  WS7_1a_Answers.docx
  WS7_1b.docx
  WS7_1b_Answers.docx

Starter suggestions

Class discussion: Think of a system that students might be familiar with, such as borrowing and returning books in a library. Lead with questions such as: A library needs to decide if the system it uses to process the borrowing and returning of books is as efficient as it could be. How can the library staff find out what is happening in the current system? If necessary, refer back to Session 6.2.

Pose further questions such as:

- Who can provide the information needed?
- How can that information be collected?
- What are the advantages and disadvantages of the methods suggested?
- How might all this information be recorded?
- How could the information be used?
- Can students begin to see some different stages or a timeline developing in their answers?

Give students a copy of **PPT7_1a** and, using the Student's Book text (pages 306–308), explain that systems have a cycle with distinct stages in which different tasks take place. The first part of this lesson will explore how information is gathered, a vital part of stage one.

Main lesson activities

Teacher explanation: Using the text on page 307, outline the four methods with simple definitions, such as:

- Interview: talking to people who are actively involved in the current system, such as the librarian, to get very specific information.

- Questionnaire: using set questions to get the opinions of as many people as possible, such as the library members, about how well a system works.

Student task: Ask students to complete **Activity 1** on page 307 of the Student's Book using **WS7_1a**.

Discussion: Review the results mentioning the advantages and disadvantages of each.

Ask students to consider whether the people to be consulted have an effect on the method chosen. For example, to find out about reaction to the ICT course, we may interview teachers and technicians but survey (put out a questionnaire to) the students. The reason for this would be the numbers of people providing information and the time required to gather that information, as well as the range of questions to be asked and the answers expected.

Give extra support by referring the students to **WS7_1a_Answers**.

Student task: Ask students to complete **Activity 2** on page 307 of the Student's Book.

Discussion: Review the results for **Activity 2**, asking students to justify their answers.

Discussion: Read through the remainder of the page text, specifically the input, processing, output table and purpose-dependent hardware and software choices.

Student task: Suggest some location relevant real-life scenarios and as a discussion or in pairs, come up with potential input, process, output and hardware, software choices.

Give extra challenge by asking students to choose a method that is suitable for the public transport investigation mentioned above. They should outline a method for carrying out the survey, describe the materials to be used and give an indication of how the completed answers will be analysed. They should focus on methods/ automation of data entry rather than statistical methods.

Plenary suggestions

WS7_1b is a chance to try and gather much of the thinking from this session. Either print out and give to students working in pairs and gather answers together or make this a class discussion drawing answers together on the board. Answers are on **WS7_1b_Answers**.

| **Assessment suggestions** | Where appropriate, the 'extra challenge' suggestion given above could be used for assessment purposes. |

7.2 The systems life cycle: Design *(double lesson)*

Learning aims
- State the advantages to a retailer of the introduction of technology, in particular EPOS and EFTPOS.
- Recognise the differences between EPOS and EFTPOS.
- Follow the symbolic representation of a system as shown in a system flowchart.
- Recognise that definition of data structure is crucial in the design phase.
- Design a suitable data input screen for a given purpose.
- Recognise the need for data validation and suggest suitable validation checks for a given scenario and associated data.

Differentiated learning outcomes
- **All students must** be able to state the advantages to a retailer of adopting technology such as POS and EFTPOS; they must be able to design a suitable data input screen for a given purpose.
- **Most students should** be able to recognise the need for defining data structure and be able to follow a system flowchart.
- **Some students could**, given the appropriate scenario, create suitable data validation checks and also be able to suggest changes to data flowcharts so as to improve the functionality of a system.

Resources
- **Student's Book:** pages 309–313
- **Files:**
  PPT7_2a
  WS7_2a
  WS7_2b
  WS7_2b_Answers
  WS7_2c
  WS7_2c_Answers

Starter suggestions

Discussion: Ask the class what they would need to know if they were tasked with designing a system that updates a supermarket stock file. Pose questions such as:

- How is data going to be stored?
- What data do you need?
- Where is the data going to come from?
- How does the data get into the system?
- What checks have to be made to ensure that the data entered is correct?

Recap with **PPT7_1a** and that the focus now is on the second stage: design.

Main lesson activities

Discussion: Work through the systems flowchart on page 311 using **PPT7_2a**. Highlight the uses of the different boxes, particularly the decision boxes that allow 'processing' to change depending on user answers or system behaviour.

Student task: Ask the students to draw a systems flowchart that details their school-day morning routine, from getting out of bed until they leave home on their way to school.

Discussion: There may be different orders for things – some will dress before breakfast and some after – but the important thing is that all students should leave the house fed and dressed!

Student task: Refer to the diagram on page 220 of the Student's Book that indicates the input, processing and output of a central heating system. Draw a system flowchart that starts with the user setting the required temperature. This flowchart will not have an ending – it will keep on going around in a circle testing the current temperature against the set temperature and switching the heater on and off.

Student task: Ask students to complete **Activity 1** on page 309, using the two columns on the left-hand side of **WS7_2a** to record their answers. Before tackling the activity, they should read the text page 305 of the Student's Book about the practical task for this unit, and also the final paragraph on page 306.

Discussion: Remind the class about EFTPOS (electronic funds transfer at point of sale). You could refer back to page 210 of the Student's Book. Ask the class the following questions:

- What differences are there between POS and EFTPOS?
- What are the advantages of EFTPOS to the retailer?
- What are the advantages of EFTPOS to the customer?
- What businesses are particularly suited to using this system?

Collect answers on the whiteboard, particularly those that refer to differences between the two systems.

Student task: Ask students to complete **Activity 2** on page 309, adding their answers into the two columns on the right-hand side of **WS7_2a**.

Discussion: Focus on the advantages of increasing the use made of technology at the checkout.

Student task: Ask students to complete **Activity 3** on page 310, using **WS7_2b** to record their answers.

Plenary suggestions

If this lesson is being delivered in two parts, you may want to have a short plenary at the end of the first session, giving a summary of what has been covered, again linking back to the systems life cycle diagram on page 306 of the Student's Book.

7.2 cont. The systems life cycle: Design (double lesson)

Learning outcomes
- State the advantages to a retailer of the introduction of technology, in particular POS and EFTPOS.
- Recognise the differences between POS and EFTPOS.
- Follow the symbolic representation of a system as shown in a system flowchart.
- Recognise that definition of data structure is crucial in the design phase.
- Design a suitable data input screen for a given purpose.
- Recognise the need for data validation and suggest suitable validation checks for a given scenario and associated data.

Differentiated learning outcomes
- **All students must** be able to state the advantages to a retailer of adopting technology such as POS and EFTPOS; they must be able to design a suitable data input screen for a given purpose.
- **Most students should** be able to recognise the need for defining data structure and be able to follow a system flowchart.
- **Some students could**, given the appropriate scenario, create suitable data validation checks and also be able to suggest changes to data flowcharts so as to improve the functionality of a system.

Resources
- **Student's Book: pages 309–313**
- **Files:**
  WS7_2b
  WS7_2b_Answers
  WS7_2c
  WS7_2c_Answers

Starter suggestions

If this part of the lesson is being delivered separately from the previous part, you may want to have a short recap before starting the main lesson activities.

Main lesson activities

Discussion: Spend some time going through the details of the design phase ('Design tasks', Student's Book page 310). Pay particular attention to the items in blue. Point out the differences between validation of data and verification of data (see 'Language boxes', Student's Book pages 310 and 312). Discuss the methods available for these processes. Point out the difficulty and cost of effective data verification. Why might the barcode not be present in the database?

Student task: Ask students to try **Activity 4** on page 311.

Discussion: Review contributions for the answers to **Activity 4**.

Discussion: Work through the second half of the session. Remind students of the need to specify type and size of data during the design of the database. The database is most likely to be a relational one, so it will also be important to consider normalisation of the data and the links between tables.

Teacher-led discussion/demonstration: Using **WS7_2b (Answers - WS7_2b_Answers)**, explain the terms: radio button, drop-down list (if necessary, refer back to Session 6.7). Highlight the necessity for the navigation buttons and the need for a search option. The student's ID and name will automatically be displayed and should be protected fields. Pressing the search button may display a dialogue box that allows you to search for records by first name, surname or ID. Additional buttons might be to go to the first or the last records.

Student task: Ask students to complete **Activity 5** on page 312. Remind students about the need for leaving enough space for the data to be entered and for the design to be 'clean', not cluttered with fields difficult to see or squashed together.

Discussion: What differences might there need to be if Mr Singh had to enter this information, not into a database, but into a paper-based system instead? He could not use drop-down boxes, for example. How else might the form change? Ask students to outline a suitable form. Could any data items be pre-printed?

Discussion: Review the table of validation checks (Student's Book page 313). Explain how they work to minimise error in data entry. Explain that these systems are not infallible, for example, the date might be in the correct format, but could still be an incorrect date.

Student task: Ask students to complete **Activity 6** (Student's Book) page 313. They could use **WS7_2c (Answers - WS7_2c_Answers)**.

Discussion: Review answers to **Activity 6**.

Give extra support by relating new learning to students' experience. For example, data validation often avoids input mistakes such as attempting to write a name in a field set as address.

Give extra challenge by asking students to set suitable validation checks that could be implemented in the design of Mr Grodzik's database.

Plenary suggestions

Remind the students of the most significant element of this stage – the file structure. Crucial to this is determining the data types of the data to be stored. This, in turn, affects the validation methods that can be used to control the data input. You can check whether students recognise this link by asking questions about methods of validation, such as: Validating username fields can be difficult because names vary so much; are there any methods by which we can limit the opportunities for error?

Assessment suggestions	Designing a suitable system flowchart often poses difficulties for students. A good method for determining progress made in this session is by setting a suitable scenario around which they are required to develop a flowchart. Choose a familiar theme such as 'leaving the house', 'making breakfast' or 'arranging to meet a friend'.

7.3 The systems life cycle: Testing and development

Learning outcomes
- Recognise the importance of suitable testing regimes prior to releasing a system for implementation.
- Suggest suitable data material for use in testing a specified system.

Differentiated learning outcomes
- **All students must** be able to suggest system tests appropriate to a given scenario.
- **Most students should** be able to describe appropriate testing methods for a given scenario.
- **Some students could** create an appropriate testing regime for a given scenario.

Resources
- **Student's Book:** pages 314–316
- **Files:**
  WS7_3a.docx
  WS7_3b.docx
  WS7_3b_Answers.docx
  WS7_3c.docx

Starter suggestions

Discussion: Ask students to read the section 'User interface' (Student's Book page 314). Give students **WS7_3a** and ask them to identify and label the input devices and output devices. Students should then write alongside these devices the data they will either collect or display. Pose the questions: How can we test that these data inputs produce the required outputs? How can we test that the sales part of this system connects properly with the stock control system? Elicit answers that point to a systematic testing programme. Bring discussion around to focus on the session aims: to test and implement a new system.

Main lesson activities

Discussion: Read and discuss the section **Testing: modules and the whole systems.** Why do the system modules need testing? What is being tested? Draw attention back to the supermarket. What modules could there be: checkout sales and stock control are the two immediate ones to consider, but these could be linked to accounting modules as well. What would the analysts at Mr Grodkik's supermarket need to consider in terms of data compatibility, hardware compatibility, memory and speed issues – some are mentioned in the text but are there any more?

Student task: Ask students to work though **Activity 1** on page 316 of the Student's Book using **WS7_3b** to record their answers.

Discussion: Ask students to consider the validation check that would be required on data that might be collected about themselves in the school student database: name (text, length), adminID (range or type), date of birth (format), tutor group (length or type or format), tutor name (text, length). What kind of extreme or abnormal data could be used to test that data?

Student task: Complete **WS7_3c**, which asks for a validation check and an abnormal item of data to be given for each data item (field).

> **Give extra support** by demonstrating data validation in an actual spreadsheet or database. Show how data can pass or be rejected by suitable validation.
>
> **Give extra challenge** by asking students to list all methods of data validation and to give examples of how they could be applied. They should show how data input could be limited by application of each of the methods and how several methods can be used to further restrict data entry. What are the advantages and disadvantages of using 'drop down boxes' (combo boxes) for data input?

Plenary suggestions

Review the topics covered. How might testing be carried out, and on what criteria for other similar systems? This could be set for homework.

| Assessment suggestions | Ask students to prepare a suitable training session for employees at Mr Grodzik's shop. |

7.4 The systems life cycle: Implementation

Learning aims
- Distinguish different implementation methods and select, with reason, a suitable method for a given scenario.
- Recognise that system change often impacts on established methods of working and actual jobs, and that staff training will be needed before implementing changes to a system.

Differentiated learning outcomes
- **All students must** be able to describe the basic process of implementation
- **Most students should** be able to state four implementation methods
- **Some students could** be able to explain multiple implementation methods.

Resources
- **Student's Book:** pages 317–318
- **Files:** WS7_4a.docx

Starter suggestions

Ask students to come up with as many different business types that in recent years will have swapped from an older IT system to a newer one as new technology has been developed. Examples to inspire students might be: Cinemas, schools or supermarkets. Discuss how they might make the transfer from one to another.

Main lesson activities

Discussion: Consider the ways of changing from one system to another (Student's Book pages 317–318), for example, your school changing from one ICT platform to another.

- Direct changeover: during the holiday period, the entire school is changed to the new platform.
- Parallel running: for an agreed timeframe, the new platform runs across the school while the old one is still being used.
- Pilot running: one area of the school gets the new platform as a test centre; when the ICT team are happy with how it works, the rest of the school is changed over.
- Phased implementation: over a period of time the new platform is gradually brought in to more and more areas of the school.

What would be the implications to you as a user for each of these changeover methods? Which would be the most likely way that your school would deal with this?

Student task: Ask students to try Activity 1 on page 317.

Discussion: Identify the possible training needs that could be relevant once the new system has been implemented. You could use the analogy of introducing a new game for friends to play. There will be a period of uncertainty for the players as they become accustomed to the new rules and procedures of the game. The period of learning can perhaps be reduced by suitable training before play is started. Training might involve going over the rules/procedures or drawing comparisons with the previous game. You could discuss methods of delivering the training, such as:

- what the medium might be (for example, computer-based, face-to-face, written)
- the nature of the training (and whether it is to be the same for all of the staff)
- at what point the training should be delivered.

Student task: Ask students to try Activity 2 on page 318.

Plenary suggestions

Go back to some of the suggested businesses or organisations in the starter and suggest a suitable implementation method for each, explaining choices.

| Assessment suggestions | Activity 4 could be carried out as an assessed task. |

7.5 The systems life cycle: Documentation and evaluation

Learning outcomes
- Distinguish between user documentation and technical documentation.
- Give appropriate reasons for differences between user and technical documentation.
- For a given system, produce elements of user and technical documentation.
- Explain the process of evaluation.
- Evaluate a piece of software.
- Recognise that the outcome of an evaluation may initiate a further cycle of system development.

Differentiated learning outcomes
- **All students must** be able to understand the need for documentation and state the differences in audience and content between user documentation and technical documentation; they should also be able to explain the need for documentation of systems and software.
- **Most students should** be able to distinguish between user and technical documentation; be able to explain the need for evaluating a new system.
- **Some students could** create detailed evaluations of a familiar system or software.

Resources
- **Student's Book:** pages 319–321
- **Files:**
  WS7_5a.docx

Starter suggestions

Obtain a number of examples of technical and user documentation. If examples cannot be brought in from home, good places to look are:

- Electrical appliance support websites
- Software support sites
- Ask school technicians for duplicate manuals for equipment or software they may have bought. Ask students about their experience of using such documents:
- How useful did they find them?
- What were the problems they encountered when using them?

Introduce the idea of the 'audience' and extend this to types of documentation needed for 'technical' and 'user' audiences. Define the terms technical and user in respect of the documentation of a system.

Student task: Ask students to sort the example documentation on display into two groups: user and technical documentation. As a supplementary task, ask them what led them to make their identification.

Main lesson activities

Discussion: Referring to the Student's Book, pages 319–320 (and the Starter above), discuss the distinctions between user and technical documentation. Draw students' attention to the 'Error handling' and 'Troubleshooting' sections of the user documentation in the table on page 320. Ask for additional contributions to the FAQs (as a prelude to starting **Activity 1**).

Student task: Ask students to work through **Activity 1** (Student's Book page 320) in pairs.

Discussion: The next section of the session is about evaluation. Depending on time available to you, it might be appropriate to evaluate a system that has some resonance with the students. You could demonstrate a popular download, streaming or online shopping site.

Ask students to evaluate the system. They will quickly realise that evaluation criteria are necessary. Now refer to the Student's Book pages 320–321 and work through the sections 'Evaluation' and 'Future developments'. Tell students that the focus for their evaluation needs to be a system with which they are familiar. Introduce **Activity 2**, asking students to think about some of the spreadsheets and databases they have created and which they could now evaluate.

Student task: Ask students to carry out **Activity 2** on page 320. Students might find it useful to use **WS7_5a** to collect their answers.

Discussion: Because **Activities 2** and **3** are so closely related, it is probably best to allow students to move to **Activity 3** (Student's Book page 320) in their own time. Reassure them that the user guide need not be complicated – a short list of bullet points is all that is necessary.

Give extra support by providing links to popular systems students can evaluate.

Give extra challenge by selecting suitable students who could tackle Activity 3 in a more interactive way by using linked webpages or as a PowerPoint presentation.

Plenary suggestions

Refer to the Starter and check that students are now much more able to distinguish between user and technical documentation.

Focus on evaluation. Ask students where in the system lifecycle it would be appropriate to run an evaluation.

Assessment suggestions	The evaluation exercises of Activities 2 and 3 could be used to assess students' progress.

7.6 Developing questionnaires and describing an existing system

Learning outcomes
- Produce a hard copy questionnaire for a particular purpose and to set criteria.
- Apply formatting to a report following set criteria.

Differentiated learning outcomes
- **All students must** be able to create a basic questionnaire that includes all the required questions and be able to format a report, although some of the formatting may be not to specification.
- **Most students should** be able to create a questionnaire and report to specification.
- **Some students could** create both items finished to professional standards, with the questionnaire printed accurately and the report sent as an email attachment to the teacher.

Resources
- **Student's Book: pages 324–326**
- **Files:**
  SystemAnalystNotes.rtf
  SystemAnalystNotesAnswers.pdf

Starter suggestions

Be aware that for **Activity 1** (Student's Book pages 324–325), printing facilities would be of benefit.

Discussion: Have a whiteboard ready for demonstration. Introduce **Activity 1** and make sure that all students are clear as to the requirements. Draw attention to the instruction that the form is to be completed 'by hand', that is, hard copy is required. Ask the students how this will affect the design process (the use of dynamic controls, such as combo boxes and check boxes, will not be possible).

Student task: Ask students to read through the list of requirements in **Activity 1** and also look carefully at the sample questionnaire (Student's Book page 324). Discuss the layout and methods for introducing clarity, such as the white space around each question.

Main lesson activities

Discussion: Question the students on what they have to do and what they can see, in general layout terms, in the sample, for example, the orientation and the navigation through the questions. Ask whether anyone can recommend any improvements.

Question the students on software choices and the benefits and drawbacks of these choices. Discuss the merits and drawbacks of using tables. If necessary, remind students about the insertion of tables and the use of cell controls, such as **Merge/Split cells** and the deletion of borders. Draw attention to the extra responses needed; discuss the effect that this might have on the layout.

Student task: Now ask students to work through **Activity 1** (Student's Book pages 324–325).

Discussion: Review progress. Be prepared to demonstrate techniques as required. Note that students will need to print their copies.

Please note: If possible, facilitate student access to a copy of the file: SystemAnalystNotes.rtf during the lesson.

Student task: Now ask students to work the remainder of the tasks on the page. Check that students are familiar with using email, and are able to attach a document and use Cc.

> Give extra support by demonstrating techniques on the whiteboard.
>
> Give extra challenge by asking students who have completed the activities to compose another version of the questionnaire, this time designed for online completion. This is covered in the next session and practice here will help in the future session.

Plenary suggestions

Review progress: Ask students to comment on how they solved problems during the session.

Assessment suggestions	Both the activities in this session offer opportunities for assessment.

7.7 Designing data capture forms and reports

Learning outcomes
- Produce output to a specified format.
- Design suitable validation methods for certain data types.
- Use formatting tools to aid in design.

Differentiated learning outcomes
- **All students must** be able to use formatting tools to produce the customer shop floor worker and receipt displays.
- **Most students should** be able to create screen displays to specification. The spreadsheet versions will show validation of text length and number size.
- **Some students could** create screen displays for the receipt that are formatted to specification and include a drop-down control.

Resources
- **Student's Book:** pages 327–330

Starter suggestions

Discussion: Find a number of images of checkouts and the hardware used – use the search terms: 'POS registers', 'Hand-held displays' or 'Hand-held stock control'.

Go through the requirements of the Quality Foods system (Student's Book page 327) and indicate to the students where the output devices mentioned in the text are found in the pictures.

Student task: Ask students to choose suitable hardware and label the parts. They should pay attention to the stated specification and ensure that the equipment meets these requirements.

Main lesson activities

Discussion: If necessary, refer again to the system needs and then to the detailed requirements for the customer display set out in **Activity 1**. You may need to revise how to use the **Left Tab** and **Right Tab** so that the name of the item and its price will index correctly within the text box. Draw attention to the instructions for sizing the text box; remember to select **Absolute** before you can enter sizes.

Student task: Ask students to work through **Activity 1** (Student's Book page 327), using the Examples from page 328 of the Student's Book to test the display.

Discussion: The next activity can be solved by again using a text box, but you may need to show the students how to use the **Centre Tab** since there are three items to be displayed in a single line.

Student task: Ask students to work through **Activity 2** (Student's Book page 328), using the examples given in the activity to test the display.

Discussion: Work through pages 328–329 of the Student's Book, introducing the idea of using a spreadsheet rather than a word processor so that data input can be validated.

Demonstration: Open the spreadsheet program that you use and demonstrate how data input to cells can be validated. Depending on the experience of the students, you may feel that it is worthwhile for them to complete the spreadsheet on page 329 of the Student's Book.

Discussion: Preview **Activity 3** making sure that the students are familiar with merging cells. Remind them that they do not need to use number as a data type unless they expect to have to perform a calculation on the data. Emphasise that the width of the checkout receipt must be no more than 8 cm. You may need to help them with this by demonstrating the use of **Page Layout** view and using the vertical and horizontal rulers to assist in sizing. It is also a good idea to set a print area since this shows as a frame.

Demonstrate: Show students how to merge cells and revise the use of the SUM function (for I15).

Student task: Ask students to work through **Activity 3** (Student's Book pages 329–330). They should use the test data shown in the Student's Book to check that the spreadsheet functions to specification.

> Give extra support by focusing on basic tasks. Students may find it much easier to size their receipt rather than add the drop-down control.
>
> Give extra challenge by introducing a '4 for 3' promotion – if customers buy four canned fruit, then they only pay for three.

Plenary suggestions

Discussion: Review progress. Ask students to comment on problems they have encountered and how they have managed to solve them.

Assessment suggestions	There are opportunities here for assessing formatting skills in both a word processor and a spreadsheet.

7.8 Developing and interpreting a model

Learning outcomes
- Implement and test a model.
- Vary data input in the model to draw conclusions.
- Use a number of rounding functions and recognise their use.

Differentiated learning outcomes
- **All students must** be able to create a basic spreadsheet in which most formulae work, with some errors.
- **Most students should** be able to make changes to their spreadsheet and draw logical conclusions from modelling using varied data.
- **Some students could** create a spreadsheet which functions to specification; draw logical conclusions from variations made to the model; and suggest changes to be made to the functionality of the model.

Resources
- **Student's Book:** pages 331–335
- **Files:** CheckoutQueue.csv

Starter suggestions

Students should set up the spreadsheet **CheckoutQueue** and save it with an appropriate file name.

Discussion: Give an overview of the spreadsheet, pointing out how it is to model the activity at the checkout and its purpose (see Student's Book page 331).

Student task: Students should enter the data in the table at the bottom of page **331** in the Student's Book, and then the data beneath the table.

Explain the action of the function ROUND, its syntax and the variations on this function. It would be helpful to the students to demonstrate this.

Main lesson activities

Discussion: Introduce **Activity 1** (Student's Book pages 332–333). The syntax of the function ROUNDUP is given, but you may want to review the use of the IF function.

Student task: Ask students to work through **Activity 1**. After making the required changes to the spreadsheet, they should check their results with the diagram on page 333.

Discussion: Review the task and clear up any difficulties. Go over the introduction to **Activity 2** (Student's Book page 333). The students need to check the accuracy of the model and it is a good idea to use simple figures to calculate for the results of changes made to the data input. If they are unable to check the maths by doing the sum in their heads, then they should use the calculator facility. Since they are making changes that could result in loss of the original data and functionality of the spreadsheet, they should save another copy, with a different filename.

Student task: Ask students to work through **Activity 2** (Student's Book page 333).

Discussion: Explain the formula that is used to calculate the percentage of customers who have to wait until the next time interval. Explain the use of the function INT and why it is used here.

Activity 3 involves replicating the functions and formulae across the spreadsheet. You may want students to save the spreadsheet with a different filename as a different version to avoid catastrophe!

Student task: Ask students to work through **Activity 3** (Student's Book page 334).

Discussion: Review the activity and the conclusions that the model indicates for Mr Grodzik. Quickly outline the changes from introduction of scanners.

Student task: Ask students to work through **Activity 4** (Student's Book page 334).

Discussion: Review the activity and the conclusions that the model indicates for Mr Grodzik's shop. Quickly outline how the model can be used to find how many more customers are required.

Student task: Ask students to work through **Activity 5** (Student's Book page 335).

Discussion: Read through the summary of recommendations given in the Student's Book on page 335.

> Give extra support by vocalising the functions in C6, C7 and C9 and 'manually' working out the value to put in the cell for each time period. Talking through the decisions that the spreadsheet is making will get the idea across in a more concrete way.
>
> Give extra challenge by introducing shift-work so that the number of tills varies throughout the day.

Plenary suggestions

Review the session, questioning where difficulties arose. Draw students' attention to the functions met and ask them why they were used in the project. Ask the students whether the model was accurate – would it reflect common practice in a convenience store? What alternatives are there for Mr Grodzik? Does the use of the model have any drawbacks? It might encourage expansion of the shop, but if changes occur to shopping habits, this could have serious consequences for Mr Grodzik because of the investment made.

Assessment suggestions	Much of the spreadsheet development is given here in detail and is therefore not useful for assessment. There are obvious ways of using similar skills but in a new scenario, such as a hotel, a petrol station or planning a bypass for a town.

8.1 Physical safety in an ICT-based environment

Learning outcomes
- Describe the potential safety dangers in a range of environments.
- Propose strategies to prevent these issues happening where possible.
- Evaluate one's own use of IT equipment to minimise potential risks.

Differentiated learning outcomes
- **All students must** be able to identify potential physical safety risks within an ICT based environment.
- **Most students should** be able to describe potential ICT physical safety risks and propose strategies to prevent them.
- **Some students could** clearly explain the vast majority of potential physical safety risks with an ICT based environment and explain detailed and practical solutions to prevent them.

Resources
- **Student's Book: pages** 340–341
- **Files:**
  PPT8_1a.pptx
  PPT8_1b.pptx
  WS8_1a.docx
  WS8_1a_Answers

Starter suggestions

Have a collection of sticky notes and ask students to take one and stick on something or somewhere in the ICT room where there is a potential danger to someone, for example plug sockets or equipment positions. Point students in the right direction if required and discuss their suggestions. Leave the notes in the same place during the lesson and return to them to see if the students' ideas were accurate.

Main lesson activities

Outline the lesson: the potential dangers of ICT and the organisations in place to educate those using it. Divide students into five groups and give each group one of the safety risks outlined in the Student's Book on page 340 and direct them to the website of the UK Health and Safety Executive or the local equivalent, if available. Each group should research content and examples related to it and then present their findings to the class. Make sure all the elements in the Student's Book have been discussed and work could be presented verbally or as an electronic presentation.

Show the photograph on **PPT8_1a Slide 1** as part of **Activity 1** on page 340 using an interactive whiteboard if possible and ask students to come and circle the potential dangers. Discuss their answers. Show and discuss **Slide 2** when ready.

Discussion: Using **WS8_1a**, ask students to apply the knowledge gained to **Activity 2** on page 341 and imagine the potential dangers in these environments. List the ICT-related equipment in each example beforehand as a basis if students need the extra support. Answers can be discussed using **WS8_1a_Answers**.

Student task: Carry out **Activity 2** and **Activity 3** on page 341. If it was possible to arrange a visitor from someone involved in health and safety as described in the Plenary then Activity 3 would best be done as homework as students could ask research questions relating to it within the discussion.

Discussion: Use **PPT8_1b** as a start point for students to quickly list equipment to add to the two areas, lots of familiar devices should be appearing but make sure time is spent considering home devices as these may bring up dangers not yet considered. If time allows students can complete **Activity 4** on page 341 or done for homework if time is short.

> Give extra support by allowing students to work in pairs. Provide lists of possible equipment from different situations. Venn diagrams could also be created for the activities to show common equipment.
>
> Give extra challenge by asking students to consider additional situations not listed, this could be done via the local health and safety organisation or website.

Plenary suggestions

If possible, ask a designated safety officer to come into the classroom and look at the potential safety issues identified at the start. Extra information and real-life examples would be really useful. If the safety officer was not available, arrange to speak with them before the lesson and identify all the potential dangers in the room. This list can then be presented to students and compared to their initial thoughts and the knowledge gained by the end of the lesson.

Assessment suggestions	Activities 3 and 4 could be written as homework and assessed. Students could also be asked to create promotional material for one of the businesses described in this session.

8.2 eSafety – *(double lesson)*

> **Learning outcomes**
> - Describe the types of personal information shared online and the potential risks.
> - Understand internet browsing habits, including social networks, shopping and gaming and strategies to protect ourselves online.
> - Describe the risks associated with email and how to minimise them.

Differentiated learning outcomes

- **All students must** be able to describe how we share personal data through the use of social networking, gaming and shopping and some of risks associated with email.
- **Most students should** be able to explain the increasing use of social networking, gaming, shopping and email and the types of personal data we share though these systems. Students should also outline strategies to protect users from any potential dangers.
- **Some students could** explain in detail how we use social networking, gaming, shopping and email and the potential dangers linked to the types of personal data we share. Each will be supported by strategies designed to minimise these risks.

Resources

- **Student's Book:** pages 342–345
- **Files:**
  PPT8_2a.pptx

Starter suggestions

Discussion: Show **PPT8_2a** and ask students to suggest pieces of data that might fall into the categories shown. This could be done using an interactive whiteboard, or projecting onto a whiteboard and using a whiteboard pen or sticking sticky notes on top of the diagram shown. Examples should include those outlined in the Student's Book on page 342 and as many others as possible. Once complete, make sure students either write down the suggestions or copy them electronically to share at a later date.

Main lesson activities

Discuss the answers provided in the starter and outline the remainder of the lesson.

Student task: Discuss **Activity 1** and ask students to work in small groups to list answers. If possible, source examples of each before the lesson, meaning they can be examined in class and students can work in small groups if required. Discuss their ideas and make sure to group similar responses together.

Discussion: Look at the list of child-related pieces of personal data on page 342 of the Student's Book and ask students to describe the potential danger that could occur from allowing access to it. Look at the Tip box on page 342 and ask students how they represent themselves online: do they use their real details, or invent details that allow a degree of anonymity? Then allow students to carry out **Activity 2** on page 343 and discuss their answers.

Discussion: Working through the section on internet browsing on page 343, make sure students understand the terminology in each of the bullet points, using real examples wherever possible. Show examples of software filters, recommended trustworthy sites and internet browser security settings on the whiteboard if possible.

Student task: Carry out **Activity 3** on page 343. Part 1 can be done as a piece of written work, possibly as homework, and part 2 should be completed as a class survey and if possible, expanded as homework, if allowing for more data, and the results collected and presented as a chart to the class.

If the double lesson is split, lesson one could end here. Students could have planned their surveys and collected data as homework.

Starter, if required:

Present and discuss results from the surveys students carried out between lessons.

Demonstration: Using a student's school email account or one of your own that you are happy to show, talk through the bullet points under the heading 'Using email' and show as many elements on-screen as possible. One way to demonstrate email is to ask students to role play being the sender, receiver, actual message and email attachment. Physically seeing the movement of students as messages, their attachments and how they can multiply bring the concept into focus for all to see and discuss.

Student task: Divide the class into three or four groups and ask them to research and present some recent large-scale email-based scams. Make sure students are clear on the terms 'spam', 'phishing' and 'smishing'. If possible, provide some starting websites to consider. Although confident in searching online, searching within a news or current affairs website can be more complex.

Discussion: Students are normally very familiar with social networking but many do not always understand the potential benefits and dangers it offers. Talk through the uses listed on page 344 and include **Activity 5** on page 344, as students should try and add additional uses not listed.

Student task: Divide the class into four groups and give each one of the four potential dangers listed on page 344. Explain to the class that they are all working for a government agency trying to promote safe internet usage. Provide each group with a large piece of paper and try and create four draft billboard or bus shelter posters that all have a similar house style but put across the dangers and safe strategies in an interesting way. If time allows, additional material such as a jingle or slogan could be created to support them.

Discussion: Internet-based gaming may have already been discussed earlier in the lesson but make sure students are aware of the dangers, and prevention strategies, to players sharing their personal information online. Ask students to complete Activity 6 and this could be given as homework.

> Give extra support by providing resources to display during the lesson and suggest news articles for students to study in Activity 4.
>
> Give extra challenge by asking students to create guides on email usage and social networking that can be shared with the class and used as revision material. These could be set as extension work or additional homework.

Plenary suggestions

Discussion: Encourage an open discussion about social networking, as described in Activity 6, the facts students have learned and any remaining questions they have. Promote the idea of students having a similar discussion at home.

Assessment suggestions	Activities 3 and 4 could be written as homework and assessed. Activity 6 could be collected and assessed.

8.3 The security of personal and commercial data

Learning outcomes
- Students should understand how data is kept secure online.
- Students will be able to define hacking and the risks to users from it.
- Students will be able to describe how computer viruses are transmitted and the problems they can cause.
- Students will develop their understanding of computer security.

Differentiated learning outcomes
- **All students must** be able to describe how websites try to keep information secret, the possible dangers we face when surfing the internet and make basic descriptions of computer security.
- **Most students should** be able to explain how websites keep information secret, the dangers we face when surfing the internet, and smart strategies to protect ourselves show an understanding of computer viruses.
- **Some students could** be able to explain in detail how websites keep information secret. They will show a clear understanding of hacking, computer viruses and security protocols. Each potential danger will be clearly matched with a protection strategy.

Resources
- **Student's Book: page** 346–349
- **Files:**
  WS8_3a.docx
  WS8_3b.docx

Starter suggestions

As students enter the room ask them to think up some passwords that might be used for the online systems they use at school or at home. Ask students not to share their real passwords but think up different ones. These could then be compared to some of the most popular used online (charts are created every year). Expand this to ask students to list the sorts of information that could be gained from accessing their password-protected systems.

Main lesson activities

Use **WS8_3a** to create ten cards and split the class into ten small groups if possible. Provide each group with a card showing a different security term or concept.

Student task: Give students 20 minutes to create a short presentation on the topic given to them on the card. Their presentation to the class should include a simple explanation of what the term means and some of the benefits and potential dangers of each.

Class Activity: Watch as many presentations as time allows and, after each one, discuss the ideas as a group, pointing out anything that differs from those listed in the Student's Book on pages 346–349. Expand on their comments as necessary. Also prompt the class to ask questions about each presentation if appropriate.

Class Activity: Students should carry out **Activity 1** and **Activity 2** (Student's Book, page 346 and 347). They will need to check what internet browser they normally use in school and test it in the classroom if possible. Discuss their answers.

Demonstration and discussion: Show and discuss examples of real-world data breach articles (the millions of hacked Yahoo accounts for example) and discuss what might happen in the future with such information being sold around the world. Expand this to discuss each of the security concerns and the strategies to prevent them.

Class activity: Take six students and provide each one with one of the speech bubbles on **WS8_3b**, as seen on page 348 of the Student's Book. Divide the rest of the class into six groups and ask the students with the card to go to one of the six groups. The student with the card should read out their dilemma and the group has a couple of minutes to offer solutions. Then the cardholder should move to another group and collect more useful tips. After the cardholder has talked to all the groups about their dilemma, they should be in a strong position to talk to the whole class about what they have learned.

> Give extra support by identifying those with additional needs, allowing them to use the Student's Book or provide some useful websites in order to carry out research.
>
> Give extra challenge by asking students to consider further incidents of hacking from around the world.

Plenary suggestions

Hold a Q&A session: What might happen if the three organisations in Activity 2 on page 347 are hacked and permission to access systems provided?

Hold a Q&A session that includes Activity 3, what developments might we see in the future to technology designed to protect our data and allow us to securely access it whenever we like?

| Assessment suggestions | Activities 2 and 3 could be set as homework if required and collected for assessment. |

8.4 Producing a word-processed report in response to a hypothesis – *(double lesson)*

Learning outcomes
- Students will understand the process involved in responding to a hypothesis.
- To be able to write a well-constructed essay response.
- Students will demonstrate formatting, proofreading and software skills that improve a document.

Differentiated learning outcomes
- **All students must** be able to identify potential physical safety risks within an ICT based environment.
- **Most students should** be able to describe potential ICT physical safety risks and propose strategies to prevent them.
- **Some students could** clearly explain the vast majority of potential physical safety risks with an ICT based environment and explain detailed and practical solutions to prevent them.

Resources
- **Student's Book:** pages 352–353
- **Files:**
PPT8_4a

Starter suggestions

Discussion: In advance of the lesson, display this hypothesis on the whiteboard: "The dangers of using technology and online systems clearly outweigh the benefits". Ask students what they think it means. Is it a balanced statement, or leaning in a certain direction? Explain to students what a hypothesis is and how, in this practical session, students will be writing a response to it as a reasoned argument that eventually agrees or disagrees with the initial statement.

Main lesson activities

Student task: Show **PPT8_4a** to the students and point out that the mind-map is unfinished. Students can either be provided with the mind-map electronically or create it from scratch but now adding as many extra 'branches' as possible in order to complete Activity 1. Provide prompting and support as required.

Student task: Before starting **Activity 2** on page 353, remind students about the importance of choosing relevant keywords in search and carefully judging the results they find, avoiding advertisements and potential fake information. Allow students 20 minutes to research their report, focusing on the elements in their mind-map they have the least knowledge of.

Student task: using their mind-map and research notes, students should write a first draft of their report as outlined in **Activity 1** on page 353.

If the double lesson was split, lesson one could end here. Students could complete their draft report as homework.

Starter if required:

Recap the practical project and discuss the five bullet points listed as key in the Report Presentation section on page 353. These should be considered in creating the final report. If students have not had their draft proofread before the lesson they should now, using a classmate. Students should be allowed to handwrite notes and make amendments to each other's work when proofreading.

Student task: As described in **Activity 4**, students should write their final report. Please note this may take an additional lesson depending on the group, or could be given as homework to complete.

> **Give extra support** by allowing students to work in pairs. Provide lists of useful websites in order to research from. Take in students draft reports and provide written feedback in addition to peer and personal proofreading.
>
> **Give extra challenge** by asking students to proofread more than one draft report and feedback to the class some of the common errors being made. A detailed bibliography of sources could be created. The report could be developed to include images, diagrams and even graphical data if it can be collected

Plenary suggestions

Hold a Q&A session: How many students agree and disagree with the hypothesis provided and why. Ask for specific examples from both sides of the argument and using the evidence collected try to decide as a class a final decision.

Assessment suggestions	The draft and final report can be collected and formally assessed.
	Students could be asked to provide evidence via screenshots on how they carried out research and also write a bibliography of their sources.